Low-Power Wide Area Network for Large Scale Internet of Things

This book presents a comprehensive exploration of LPWANs, delving into their fundamental concepts, underlying technologies, and the multifaceted challenges they tackle. This book recognizes that LPWANs don't operate in isolation; they are intimately intertwined with Artificial Intelligence and Machine Learning (AI/ML) technologies, which play a pivotal role in optimizing LPWAN performance and capabilities. The book is a collection of original contributions regarding air interface, transmission technologies and novel network architectures, such as network slicing, cloud/fog/edge computing, ad hoc networks and software-defined network. Also, this book provides a guide for researchers of IoT applications to choose suitable LPWAN technologies and describe the design aspects, network architectures, security issues and challenges.

Features:

- Explains machine learning algorithms onto low-power wide area network sensors for compressed communications.
- Illustrates wireless-based Internet of Things networks using low-power wide area networks technology for quality air.
- Presents cognitive Internet of Things networks using wireless communication and low-power wide area network technologies for Ad Hoc networks.
- Discusses a comprehensive study of low-power wide area networks for flying Ad Hoc networks.
- Showcases the study of energy efficient techniques aided by low-power wide area network technologies for the Internet of Things networks.

The text is aimed at senior undergraduate, graduate students, and academic researchers in the fields of electrical engineering, electronics and communication engineering, computer engineering, and information technology.

Prospects in Networking and Communication
Series Editor: Mohammad M. Banat

Low-Power Wide Area Network for Large Scale Internet of Things: Architectures, Communication Protocols, and Recent Trends
Mariyam Ouaissa, Mariya Ouaissa, Inam Ullah Khan, Zakaria Boulouard, and Junaid Rashid

Internet of Things: A Hardware Development Perspective
Ayoub Khan

For more information about this series, please visit: https://www.routledge.com/ Prospects-in-Networking-and-Communications/book-series/NETCOM

Low-Power Wide Area Network for Large Scale Internet of Things

Architectures, Communication Protocols and Recent Trends

Edited by
Mariyam Ouaissa, Mariya Ouaissa,
Inam Ullah Khan, Zakaria Boulouard and
Junaid Rashid

CRC Press
Taylor & Francis Group
Boca Raton London New York

CRC Press is an imprint of the
Taylor & Francis Group, an **informa** business

Front cover image: Profit_Image/shutterstock

First edition published 2024
by CRC Press
2385 NW Executive Center Drive, Suite 320, Boca Raton FL 33431

and by CRC Press
4 Park Square, Milton Park, Abingdon, Oxon, OX14 4RN

CRC Press is an imprint of Taylor & Francis Group, LLC

ISBN: 978-1-032-53079-6 (hbk)
ISBN: 978-1-032-73165-0 (pbk)
ISBN: 978-1-003-42697-4 (ebk)

DOI: 10.1201/9781003426974

Typeset in Sabon
by codeMantra

Contents

Preface

The Internet of Things (IoT) is a phenomenon that's reshaping our world, connecting countless devices, and driving innovation across industries. However, IoT's diverse applications come with unique communication challenges, especially regarding long-distance, low-power connectivity. It's here that Low Power Wide Area Networks (LPWANs) emerge as a crucial solution, complementing traditional wireless technologies by efficiently serving the connectivity needs of IoT and Machine-to-Machine (M2M) devices.

In our edited book, *Low-Power Wide Area Network for Large Scale IoT: Architectures, Communication Protocols and Recent Trends*, we embark on a comprehensive exploration of LPWANs, delving into their fundamental concepts, underlying technologies, and the multifaceted challenges they tackle. This book recognizes that LPWANs don't operate in isolation; they are intimately intertwined with Artificial Intelligence and Machine Learning (AI/ML) technologies, which play a pivotal role in optimizing LPWAN performance and capabilities.

Let's take a closer look at the specific themes and contributions of each chapter:

- The first and foundational chapter provides a deep dive into LPWAN technology, offering insights into its core principles, enabling technologies, and the challenges that come with deploying large-scale IoT systems.
- Chapter 2 explores the synergy between LPWANs and Machine Learning, showcasing how ML techniques can enhance communication reliability, efficiency, and intelligence within IoT networks.
- In Chapter 3, we venture into real-world applications of LPWANs powered by Machine Learning, illustrating how these technologies enable innovative solutions across various domains.
- Chapter 4 conducts an in-depth study, focusing on the symbiotic relationship between IoT networks, Machine Learning, and LPWAN technology, elucidating how they influence each other.

- Chapter 5 reviews the interplay between LPWANs and cellular networks, offering insights into how these technologies can collaborate to meet the demands of IoT applications.
- Chapter 6 investigates how LPWANs can be applied to demand-side management, optimizing resource consumption and enhancing energy efficiency.
- Focusing on air quality monitoring, Chapter 7 explores LPWANs empowered by AI, which play a vital role in ensuring the quality of the air we breathe.
- Chapter 8 conducts a thorough review of AI techniques, illustrating how they elevate the efficiency and intelligence of LPWAN sensor networks.
- Chapter 9 dives into the world of Software-Defined Networking (SDN) and cloud computing, elucidating their role in securing LPWAN-based IoT systems.
- Chapter 10 looks ahead to 6G and explores how big data analytics will enable massive IoT deployments, leveraging LPWAN technology.
- In the context of healthcare, Chapter 11 showcases the role of LPWAN networks in revolutionizing patient care and health monitoring.
- Chapter 12 provides a specific use case of LPWAN, focusing on LoRa networks, in revolutionizing smart healthcare applications.
- The final chapter delves into the critical topic of early detection of diseases, particularly COVID-19, using LPWAN networks and deep learning techniques.

Each chapter serves as a piece of the larger puzzle, contributing valuable insights, innovations, and solutions to the complex landscape of LPWANs for IoT. We invite you to journey through this book, exploring the cutting-edge developments, challenges, and promising possibilities that await. As we collectively shape the future of IoT and LPWANs, may this book serve as a beacon, guiding us toward a more connected, intelligent, and sustainable world.

About the editors

Dr. Mariyam Ouaissa is currently an Assistant Professor in Networks and Systems at ENSA, Chouaib Doukkali University, El Jadida, Morocco. She received her Ph.D. in 2019 from the National Graduate School of Arts and Crafts, Meknes, Morocco and her Engineering Degree in 2013 from the National School of Applied Sciences, Khouribga, Morocco. She is a communication and networking researcher and practitioner with industry and academic experience. Dr Ouaissa's research is multi-disciplinary that focuses on Internet of Things, M2M, WSN, vehicular communications and cellular networks, security networks, congestion overload problem and the resource allocation management and access control. She is serving as a reviewer for international journals and conferences including *IEEE Access, Wireless Communications and Mobile Computing.* Since 2020, she has been a member of "International Association of Engineers IAENG" and "International Association of Online Engineering", and since 2021, she is an "ACM Professional Member". She has published more than 30 research papers (this includes book chapters, peer-reviewed journal articles, and peer-reviewed conference manuscripts), 10 edited books, and 6 special issues as a guest editor. She has served on Program Committees and Organizing Committees of several conferences and events and has organized many Symposiums/Workshops/Conferences as a General Chair.

Dr. Mariya Ouaissa is currently a Professor in Cybersecurity and Networks at Faculty of Sciences Semlalia, Cadi Ayyad University, Marrakech, Morocco. In 2019, she received her Ph.D. in Computer Science and Networks at the Laboratory of Modelisation of Mathematics and Computer Science from ENSAM-Moulay Ismail University, Meknes, Morocco. She is a Networks and Telecoms Engineer, graduated in 2013 from the National School of Applied Sciences Khouribga, Morocco. She is a Co-Founder and IT Consultant at IT Support and Consulting Center. She worked for the School of Technology of Meknes Morocco as a Visiting Professor from 2013 to 2021. She is a member of the International Association of Engineers and International Association of

Online Engineering, and since 2021, she has been an "ACM Professional Member". She is an expert reviewer with Academic Exchange Information Centre (AEIC) and Brand Ambassador with Bentham Science. She has served and continues to serve on technical program and organizer committees of several conferences and events and has organized many Symposiums/Workshops/Conferences as a General Chair and also as a reviewer of numerous international journals. Dr. Ouaissa has made contributions in the fields of information security and privacy, Internet of Things security, and wireless and constrained networks security. Her main research topics are IoT, M2M, D2D, WSN, Cellular Networks, and Vehicular Networks. She has published over 40 papers (book chapters, international journals, and conferences/workshops), 12 edited books, and 8 special issues as a guest editor.

Inam Ullah Khan is the Founder of Internet of Flying Vehicles-Lab at AI-EYS. Recently, he has been working as a global mentor/guest lecturer at Impact Xcelerator, IE School of Science and Technology, Madrid, Spain. Previously, he was working as a visiting researcher at King's College London, United Kingdom. Also, he was a faculty member at different universities in Pakistan including Center for Emerging Sciences Engineering & Technology (CESET), Islamabad, Abdul Wali Khan University, Garden Campus, Timergara Campus, University of Swat & Shaheed Zulfikar Ali Bhutto Institute of Science and Technology (SZABIST), Islamabad Campus. He completed his Ph.D. in Electronics Engineering from the Department of Electronic Engineering, Isra University, Islamabad Campus, School of Engineering & Applied Sciences (SEAS). Also, he completed his M.S. in Electronic Engineering at the Department of Electronic Engineering, Isra University, Islamabad Campus, School of Engineering & Applied Sciences (SEAS). He had done undergraduate degree in Bachelor of Computer Science from Abdul Wali Khan University Mardan, Pakistan. He authored/coauthored more than 50 research articles in reputable journals, conferences, and book chapters. His research interest includes Network System Security, Intrusion Detection, Intrusion Prevention, cryptography, Optimization techniques, WSN, IoT, Mobile Ad Hoc Networks (MANETS), Flying Ad Hoc Networks, and Machine Learning. He served in many international conferences as Session Chair/Technical program committee member. Also, he served as a guest editor with many prestigious international journals. Apart from that, he is the General Chair at the International Conference on Trends and Innovations in Smart Technologies (ICTIST'22), virtually from London, United Kingdom. In addition, he also served as an editor of several books. More interestingly, he was invited as a technology expert many times on Pakistan National Television and other media outlets.

Dr. Zakaria Boulouard is currently a Professor at the Department of Computer Sciences at the "Faculty of Sciences and Techniques Mohammedia, Hassan

II University, Casablanca, Morocco". In 2018, he joined the "Advanced Smart Systems" Research Team at the "Computer Sciences Laboratory of Mohammedia". He received his PhD in 2018 from "Ibn Zohr University, Morocco" and his Engineering Degree in 2013 from the "National School of Applied Sciences, Khouribga, Morocco". His research interests include Artificial Intelligence, Big Data Visualization and Analytics, Optimization and Competitive Intelligence. Since 2017, he has been a member of "Draa-Tafilalet Foundation of Experts and Researchers", and since 2020, he has been an "ACM Professional Member". He has served on Program Committees and Organizing Committees of several conferences and events and has organized many Symposiums/Workshops/ Conferences as a General Chair. He has served and continues to serve as a reviewer of numerous international conferences and journals. He has published several research papers. This includes book chapters, peer-reviewed journal articles, peer-reviewed conference manuscripts, edited books, and special issue journals.

Dr. Junaid Rashid is a Postdoctoral Researcher at the Department of Computer Science and Engineering, Kongju National University, South Korea, from August 2021. He worked as an Honorary Senior Postdoctoral Research Fellow at the Center for AI and Data Science, Edinburgh Napier University, Scotland, UK, since November 2020 and was awarded the Fellowship in March 2022. In 2020, he received his Ph.D. from the Department of Computer Science, University of Engineering and Technology, Taxila, Pakistan. He also received his M.S and B.S in Computer Science from COMSATS Institute of Information Technology, Wah Campus, Pakistan in 2016 and 2014, respectively. He worked as a lecturer in the Department of Computer Science at AIR University Islamabad, Kamra Campus, Pakistan, from January 2020 to July 2021. He served as a research associate and lecturer from March 2017 to January 2020 at the University of Engineering and Technology, Taxila, Pakistan. He has served as a lecturer (Visiting) from September 2019 to February 2020, September 2016 to January 2017 and September 2015 to January 2016 in COMSATS University Islamabad, Wah, Pakistan. He received a fully funded scholarship for his Ph.D. degree. He has been working as an academic editor in PLOS One. He is a program committee member at the International Conference on Advances in Communication Technology and Computer Engineering (ICACTCE'22), Morocco. He has served/been serving as a reviewer of various reputed journals and conferences. He has published many research papers in prestigious journals and conferences. His research interest includes Machine Learning, Data Science, Natural Language Processing, Topic Modeling, Text Mining, Information Retrieval, Pattern Recognition, Fuzzy Systems, Medical Informatics, Computer Networks, and Software Engineering.

Contributors

Bommidala Abhishake
Dr. M.G.R Educational and
    Research Institute
Chennai, India

Abdul Ahad Abro
Ege University
Bornova, Türkiye

Samir Ahmad
MTN Consulting LLC.
Chandler, Arizona

Abdullah Akbar
National University of Computer
    and Emerging Sciences
Lahore, Pakistan

Rehan Akbar
Universiti Teknologi PETRONAS
Perak, Malaysia

Qasem Abu Al-Haija
Princess Sumaya University for
    Technology
Amman, Jordan

Nasreen Anjum
University of Gloucestershire
Gloucestershir, United Kingdom

Muhammad Yaseen Ayub
COMSATS University Islamabad
Attock, Pakistan

Zakaria Boulouard
LIM, Hassan II University
Casablanca, Morocco

L. Dharani
Dr. M.G.R Educational and
    Research Institute
Chennai, India

Sarah El Himer
Electrical Engineering Department
Sidi Mohammed Ben Abdellah
    University
Fez, Morocco

G. Victo Sudha George
Dr. M.G.R Educational and
    Research Institute
Chennai, India

Najia Ait Hammou
Faculty of Science Semlalia
Cadi Ayyad University
Marrakech, Morocco

Billa Hemalatha
Dr. M.G.R Educational and
   Research Institute
Chennai, India

Waheeduddin Hyder
Millennium Institute of Technology
   & Entrepreneurship (MiTE)
Karachi, Pakistan

Aftab Alam Janisar
Universiti Teknologi PETRONAS
Perak, Malaysia

Awais Khan Jumani
School of Electronic and
   Information Engineering
South China University of
   Technology
Guangdong, China
and
ILMA University
Karachi, Pakistan

Sapna Juneja
KIET Group of Institutions
Delhi NCR, India

Samina Kanwal
COMSATS University Islamabad
Rawalpindi, Pakistan

Samira Kanwal
University of Engineering and
   Technology
Taxila, Pakistan

Siva Kumar Gowda Katta
Sreenidhi Institute of Science and
   Technology
Hyderabad, India

Maysa Khalil
Princess Sumaya University for
   Technology
Amman, Jordan

Inam Ullah Khan
Isra University
Islamabad, Pakistan

Ifrah Komal
Riphah International University
Islamabad, Pakistan

Nalli. Vinaya Kumari
Malla Reddy Institute of
   Technology & Science
Hyderabad, India

Brahim Lakssir
Digitalization and Microelectronis
   Smart Devices Department
MAScIRv
Rabat, Morocco

Hajar Mousannif
Faculty of Science
Semlalia Cadi Ayyad University
Marrakech, Morocco

Zaid Muhammad
Abdul Wali Khan University
Mardan, Pakistan

Shagufta Mujeeb
Dr. Ruth K. M. Pfau, Civil Hospital
Karachi, Pakistan

Farhood Nishat
Institute of System Engineering
Riphah University
Islamabad, Pakistan

Mariya Ouaissa
Computer Systems Engineering
    Laboratory
Cadi Ayyad University
Marrakech, Morocco

Mariyam Ouaissa
Laboratory of Information
    Technologies
Chouaib Doukkali University
El Jadida, Morocco

Lokaiah Pullagura
Jain University
Bangalore, India

Muhammad Allah Rakha
FAST National University of
    Computer and Emerging
    Sciences
Peshawar, Pakistan

Junaid Rashid
Sejong University
Seoul, Republic of Korea

Mujeeb ur Rehman
Universiti Teknologi PETRONAS
Perak, Malaysia

Abdul Qahar Shahzad
Quaid-i-Azam University
Islamabad, Pakistan

Waqas Ahmed Siddique
Millennium Institute of Technology
    & Entrepreneurship (MiTE)
Karachi, Pakistan

Muhammad Farhan Siddiqui
Karachi Institute of Economics and
    Technology (KIET)
Karachi, Pakistan

Muhammad Arabi Tayyab
Karachi Institute of Economics and
    Technology (KIET)
Karachi, Pakistan

Ubaid Ullah
University of Wah
Wah, Pakistan

Muhammad Usama
University of Wah
Wah, Pakistan

Zupash
COMSATS University Islamabad
Attock, Pakistan

Chapter I

Low-power wide area network for large-scale IoT

Fundamentals, technologies, and challenges

Mariyam Ouaissa
Chouaib Doukkali University

Mariya Ouaissa
Cadi Ayyad University

Zakaria Boulouard
Hassan II University

Sarah El Himer
Sidi Mohammed Ben Abdellah University

Inam Ullah Khan
Isra University

1.1 INTRODUCTION

Low Power Wide Area Network (LPWAN) represents a new communication paradigm, complementing traditional wireless and short-range cellular technologies [1]. The aim of these networks is to meet various requirements of Internet of Things (IoT) applications. LPWAN technology is distinguished from other wireless technologies by its extensive connectivity for terminals with low power consumption and low data rate. This feature makes them popular: around a quarter of all 30 billion IoT devices by 2025 will need to be connected to the Internet through wireless radio networks. There are various applications in several industries using LPWAN technologies. These industries include, among others, the smart city, personal IoT applications, smart grid, smart metering, industrial monitoring, and agriculture [2].

LPWAN networks are unique; in that, they provide trade-offs between different technologies used in the context of IoT, such as short-range wireless networks (ZigBee, Bluetooth, Z-Wave, WLAN, and Wi-Fi) and networks

DOI: 10.1201/9781003426974-1

I

traditional cellular (GSM, UMTS, and LTE). Indeed, the former are not ideal for connecting terminals with low consumption, distributed over large geographical areas. The range of these technologies is limited to a few hundred meters at best. Therefore, terminals cannot be deployed or moved arbitrarily anywhere, which is necessary for many applications in cities and territories. To remedy this, a dense deployment of terminals and the use of gateways connected through a multi-hop mesh network are necessary, with an impact on deployment costs. WLANs are characterized by shorter coverage areas and higher power consumption for Machine to Machine (M2M). As for the second- and third-generation cellular networks, they provide more extensive coverage and are more suitable for M2M. However, the unlicensed use of the frequency bands assigned to certain standards has made it possible to broaden the field of low-power technologies. In general, traditional cellular technologies do not achieve sufficient energy efficiency to provide battery life for a few years. The complexity and cost of cellular terminals are high due to their ability to process complex waveforms to cope with throughput, mobility, and location [3].

LPWANs are not designed to handle all IoT use cases, but rather intended for applications that tolerate delays and low speeds, and generally require low power consumption and low cost. This is the reason why several LPWAN standards have emerged in recent years. LPWAN are to be separated into two categories: cellular and non-cellular networks. The non-cellular network, dedicated to the IoT, is mainly composed of proprietary networks while the cellular category is based on existing networks [4].

The objective of this chapter is to provide a guide for researchers of IoT applications to choose suitable LPWAN technologies. Thus, we focus on the comparison of several technologies, Ingenu, Weightless W, N, P, Lora, Sigfox, NB-IoT, and LTE-M which are promising technologies in terms of service demand, cost, and power efficiency.

The rest of this chapter is organized as follows: Section 1.2 discusses LPWAN characteristics and applications. In Section 1.3, we describe the design considerations of LPWAN. A technical comparison between several LPWAN technologies is addressed in Section 1.4. Section 1.5 presents the challenges faced by LPWAN technologies. Finally, we conclude in Section 1.6.

1.2 OVERVIEW OF LPWAN

In this section, we present the architectures and applications area of LPWAN networks.

1.2.1 Presentation of LPWAN

One of the primary issues for IoT and M2M is long-distance and low-power connectivity. LPWAN network is built on a central hub that controls all connections. As a result, each sensor will be able to send messages to a local hub but not directly to other transmitters. This technology allows for the transmission and reception of very short messages, with the added benefit that the components used to deliver these signals are both inexpensive and energy-efficient. It is now possible to send a few messages per day for ten years using a simple battery [5].

LPWAN networks are also distinct from standard short-range networks such as ZigBee, Bluetooth, and Wireless Local Area Network (WLAN). A tighter mesh of gateways or a mesh architecture can occasionally compensate for the range of these communicating devices, which peaks out at roughly one hundred meters. As a result, network infrastructure installation and maintenance costs will be substantially higher. The use cases appear to be more similar to the traditional uses of Global System for Mobile Communication (GSM) and Long-Term Evolution (LTE). Despite their extensive coverage, GSM networks do not compete in the critical areas of IoT, particularly consumption and pricing [6].

Figure 1.1 depicts where the LPWAN networks are in relation to one another in terms of power consumption, flow, and distance. It is worth noting that they occupy a crucial position on the graph that is not covered by any other communication mechanism.

1.2.2 LPWAN application areas

Low-power and wide-coverage networks represent a new revolution in the world of wireless communication, enriching and complementing cellular technologies and short-range wireless networks [7]. LPWAN technology allows extensive connectivity for terminals with low power consumption

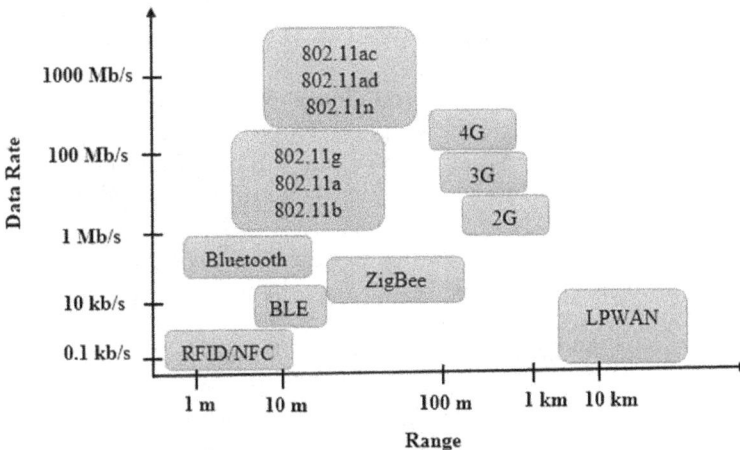

Figure 1.1 Data rate vs. range: LPWAN positioning.

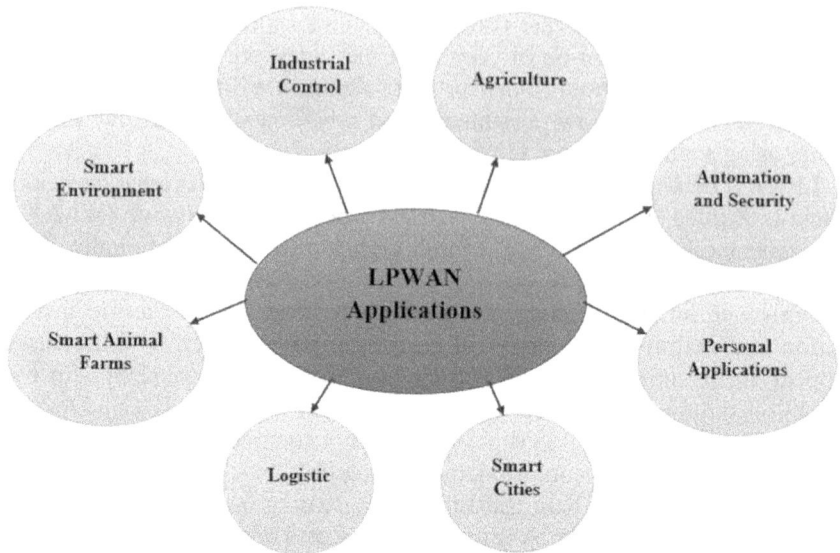

Figure 1.2 LPWAN application areas.

and low data rate. Figure 1.2 illustrates different business sectors using the application of LPWAN technologies [8]. These sectors of activity include, among others:

1. **Smart environment:** This is about preventing accidents and natural disasters. One can imagine the possibility of detecting fires, measuring air pollution, or for example the snow level. It can be used to prevent avalanches, floods, and drought. Detecting earthquakes is also a possibility.
2. **Smart cities:** Smart parking, sound monitoring, people detection, traffic management, public lighting, etc.
3. **Smart agriculture:** Monitoring of vines, monitoring of greenhouses, control of land irrigation, weather station, monitoring of compost, tracking of animals, etc.
4. **Smart animal farms:** Livestock traceability, toxic gas level monitoring, animal development monitoring, hydroponic crop monitoring, etc.
5. **Logistics:** Monitoring of transport conditions, location of packages, detection of storage incompatibility, fleet traceability, etc.
6. **Industrial control:** Machine monitoring, equipment condition, indoor air quality, temperature monitoring, ozone level detection, equipment location, vehicle diagnostics, etc.

1.3 DESIGN CONSIDERATIONS OF LPWAN NETWORKS

The success of LPWAN technologies is based on their ability to provide low-power connectivity to a large number of terminals spread across large geographic areas at a low cost. This section describes the techniques used by LPWAN technologies to achieve these often contradictory goals. We emphasize that LPWAN technologies share some design goals with other wireless technologies. The main goal of LPWAN technologies, however, is to achieve wide reach with low power consumption and low cost, unlike other technologies where it may be more important to achieve higher throughput, lower latency, and increased reliability [9,10].

1.3.1 Coverage

LPWAN technologies are designed to cover a large area and allow good signal propagation to hard-to-reach indoor locations. Quantitatively, a gain of +20 dB compared to existing cellular systems is targeted. Terminals can link to gateways from a few tens of kilometers away, depending on the deployment environment. To accomplish this goal, the sub 1 GHz spectrum and the unique modulation methods described below are used.

- **Frequency usage below 1 GHz:** Most LPWAN technologies use carrier frequencies below 1 GHz, which provides robust and reliable communication with low power budgets. The sub 1 GHz band is less congested than the 2.4 GHz used by most wireless technologies so low frequency signals experience less attenuation and multipath fainting caused by obstacles.
- **Modulation techniques:** LPWAN technology is designed to achieve a link budget of 150±10 dB, which allows a range of a few tens of kilometers in urban and rural areas, respectively. The physical layer conditions a high data rate and slows down the rate of modulation in order to put more energy into each bit or symbol. For this reason, receivers can correctly decode heavily attenuated signals. Two types of modulation techniques adapted by LPWAN technologies: narrow band technique and spread spectrum technique.

1.3.2 The consumption

To capitalize on the massive commercial opportunity presented by battery-powered IoT terminals, ultra-low power consumption operation is a must [11]. To reduce maintenance costs, AA or button cell batteries with a long life of ten years or more are preferred. Low consumption depends on the following parameters:

- **Network topology**: Many current networks are based on a mesh structure. Each point of the network is therefore capable of relaying information to send it to the concentrator. Some critical paths may lead to certain modules being used more than others. Overall, autonomy is therefore revised downwards. In the case of LPWANs, the star topology implies a direct connection between each element of the network and the hub. There is, therefore, no need for devices to remain on permanently to relay hypothetical messages.
- **Duty cycle**: Low consumption can be achieved by switching off certain energy-consuming components of IoT devices such as radio transmitters. In the case of LPWANs, the duty cycle makes it possible to cut the transmitters and to switch them on only when data must be sent or received. When it comes to receiving data from the hub, different methods can be implemented, in particular the planning of the opening of time windows during which the objects remain on. In this case, it will be necessary to configure the device as soon as it joins the network. Finally, in order to regulate the data exchanges as a function of the bands used, a speaking time per device is imposed which does not exceed 1% of the time over one hour.
- **ALOHAnet protocol**: Extremely simple to set up, this protocol consists of sending data packets to a transmitter as soon as he wishes. If any data is corrupted or lost, this packet will be retransmitted after a random waiting time. This scheme is therefore much less energy-intensive than the conventional Medium Access Control (MAC) protocols used for cellular networks.

All these elements, allowing to considerably reduce the consumption of the modules, are essential to lower maintenance costs significantly.

1.3.3 The costs

The low-cost design of terminals is made possible by several techniques, summarized in the following.

- **Reduced complexity of terminal hardware**: Unlike short-range cellular and wireless technologies, transceivers LPWAN must deal with less complex waveforms. This allows them to reduce data rates and memory size while minimizing hardware complexity.
- **Minimal infrastructure**: LPWAN technologies use a single gateway that can connect tens of thousands of devices over several kilometers, significantly reducing costs for network operators.

I.3.4 Scalability

The use of a large number of terminals exchanging low volumes of traffic is one of the main requirements of LPWAN technologies. Several techniques are considered to ensure this operation:

- **Diversity:** To operate multiple connected devices, it is essential to efficiently exploit the diversity of channels, time, and hardware. LPWAN technologies use multi-channel and multi-antenna communications. Multi-channel communications are more resistant to interference, whereas multi-antenna communications are used to parallel transmissions to and from connected terminals.
- **Densification:** LPWANs and traditional cellular networks use a dense deployment of gateways to cope with increasing terminal density in some areas. The problem is to do this without interference between terminals and gateways. Because conventional cellular techniques rely on well-coordinated management of radio resources within and between cells, new densification approaches for LPWAN networks require additional research because most LPWAN systems do not.
- **Adaptive channel selection and data rate:** The optimization of the links for reliable and energetic communication in the LPWAN networks depends on several techniques: the adaptation of the modulation schemes, selection of the best channels to ensure the range or to make an adaptive transmission power control which requires an effective monitoring of the link qualities and coordination between the terminals and the network. The possibility of adaptive channel selection and modulation depends on each LPWAN technology. In the event that the gateway cannot provide feedback on the quality of the communication on the uplink, these use a very simple mechanism to improve the quality of the link: the repeated transmission of the same packet several times, often on several randomly selected channels, in the hope that at least one copy reaches the gateway successfully. These mechanisms help improve uplink reliability while keeping the complexity and cost of terminals very low. In cases where downlink communication can allow adaptation of uplink parameters, gateways or the core of the network can play a critical role in selecting parameters such as the optimal channel or throughput to improve reliability and energy efficiency.

I.3.5 Quality of service

LPWAN technologies are aimed at a wide range of applications with variable Quality of Service (QoS) needs. Some applications do not require it such as smart meter, while others performance is highly dependent such as alarms generated by home security applications. For cellular standards where radio resources can be shared between LPWAN and mobile broad-band applications, mechanisms will need to be defined for the coexistence of different throughput levels.

When users of LPWAN technologies talk about QoS, they generally refer to particular requirements for message delivery, such as time of arrival, probability of successful reception, and rate of packet rejection. Most users understand that there are trade-offs related to QoS. For example, message transmission delay is often caused by network congestion. A simple oversizing of the network can therefore improve the quality of service. Short delay times generally limit the ability of devices to go into sleep mode and therefore reduce battery life. The best approach is to set up a flexible system that can achieve different QoS results as needed, prioritizing high QoS traffic such as alarms according to lower traffic.

The orders of magnitude of range, speed and autonomy are specific to this type of network. However, it is impossible to maximize these quantities with a single network. This is why there is a multitude of LPWAN technologies that allow compromises depending on the problems of the application.

1.4 TECHNICAL COMPARISON OF LPWAN TECHNOLOGIES

In full emergence, LPWAN networks are to be separated into two categories. These are cellular (NB-IoT, LTE-M)) and non-cellular (Ingenu, Weightless W, N, P, Lora, and Sigfox) networks. The latter, dedicated to IoT, is mainly made up of proprietary networks while the cellular category is based on existing networks (Figure 1.3) [12,13].

1.4.1 Ingenu

It is the proprietary network deployed by Ingenu that offers two-way communication. Based on Random Phase Multiple Access (RPMA), this network is one of the few to operate on frequencies above GHz, namely 2.4 GHz. The use of this band, not subject to duty cycle regulations in Europe and the United States, implies a high speed for this type of communication. The values reached are of the order of a hundred kbps. Conversely, consumption is higher than for other technologies and the range, estimated at around 5 kilometers, is one of the lowest in the sector (Ingenu RPMA).

1.4.2 Weightless W, N, P

Three standards developed by the Weightless Special Interest Group set up on Sub-1GHz frequency bands and on free bands previously allocated to terrestrial television broadcasting (WEIGHTLESS).

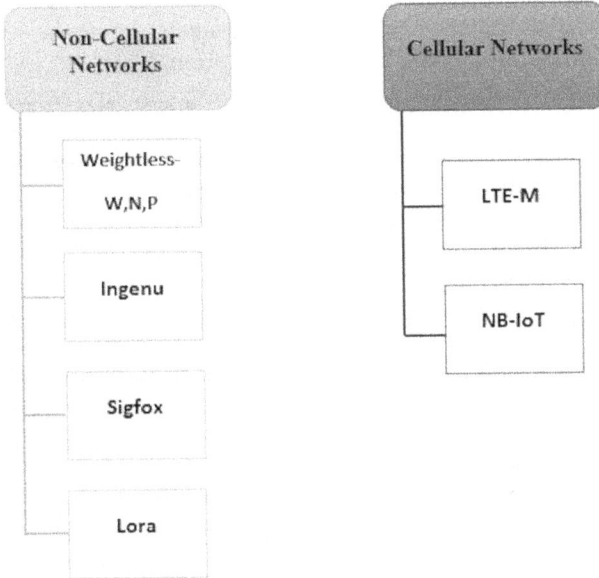

Figure 1.3 Major emerging technologies LPWAN.

In general, each standard has distinct characteristics in order to provide solutions for the three concepts: range, speed, and autonomy. Cost remains an essential factor; however, it is Weightless-N that tends to minimize this factor. This standard, operating in the Ultra High Frequency (UHF) 800–900 MHz band and based on Ultra Narrow Band (UNB) technology, only works for uplink communications. It is therefore impossible, for the server, to send data to the communicating nodes. The consumption of Weightless-N allows one to reach autonomy of up to ten years while having a relatively large range. In addition, considering the throughput, which is in the average of LPWAN reaching up to 100 kbps and low overall costs, the Weighless-N is an ideal candidate to build a network quickly and at low cost. On the other hand, its inability to offer bottom-up communication significantly reduces its scope.

The Weightless-W is, therefore, a good alternative if the use case attaches particular importance to two-way communication. It operates in a lower frequency band, between 470 and 790 MHz. This factor, coupled with a narrowband architecture, makes it possible to multiply a maximum throughput by 10 while sacrificing the range and increasing the consumption of the communicating system.

Offered as a high-performance means guaranteeing a high level of quality of bidirectional LPWAN communication service, the Weightless-P can operate on a panel of frequencies such as 169, 433, 470, 780, 868, 915, and 923 MHz. Despite speed, autonomy, and range at half-mast compared to other Weightless Group standards, the built-in functionality of this network allows users to set up much more complex systems. Thus, the Weightless-P offers data encryption in Advanced Encryption Standard (AES) or, for example, to automatically manage the throughput by optimizing radio resources according to the quality and strength of the signal. This latter standard is, therefore, useful in complex applications requiring advanced functionalities. In the future, it could therefore be of interest to couple these different standards to offer bidirectional communications or to optimize certain parameters.

1.4.3 Sigfox

Sigfox is a French Company, which operates an LPWAN network and acts as an access provider. Most of the information available comes from sales presentations from Sigfox. To date, the network is particularly developed in Europe. The technology used by Sigfox is based on UNB, a proprietary solution in the 868 MHz free band [14]. The carrier used by transmissions is 100 Hz. The use of such a thin band limits the symbol rate to 100 baud, but greatly increases the link budget, so that a large coverage is provided without the use of coding. To ensure the demodulation of a message without the presence of interferers, the messages are repeated three times, on three different center frequencies considered as random. The use of the tape free limits the maximum number of messages per terminal to 14 per day. A downlink from the central node to the terminals is available (SIGFOX).

1.4.4 LoRa

Long Range (LoRa) is a technology based on the use of a patented modulation based on frequency spreading by Chirp Spread Spectrum (CSS). Messages can be transmitted at different spreading factors noted Spreading Factor (SF), ranging from 7 to 12, and on different carrier sizes, ranging from 125 to 500 kHz. A LoRa symbol, the duration of which is inversely proportional to the carrier used, is represented by a continuous phase signal whose frequency varies linearly in the subcarrier considered [15]. One bit of information is carried by SF LoRa symbols; the chosen SF value is large, whereas the spectral efficiency will be low. This reduces throughput and makes transmissions more robust to interference. In addition, the modulation uses a coding whose rate can vary between 1/2 and 4/5. A link from the central node to the terminal is possible; a reception window is opened by the terminal after each transmission. This modulation is coupled with the LoRaWAN [16], open source communication protocol, promoted by the LoRa Alliance, an open consortium that seeks to standardize the use of the technology. Each user can administer their own LoRa network and make it compatible with that of other users, provided they purchase specific terminals (LoRa Alliance).

1.4.5 NB-IoT

Arrived on the market much later than the technologies previously described, Narrowband IoT (NB-IoT) was proposed by the 3GPP ((3rd Generation Partnership Project) standardization body in its version 13 as an extension of LTE in June 2016. It's a standard that will be used by existing operators on regulated frequency bands. To date, the network is operational in several European countries, as well as in Brazil, China, and the United States. This standard is based on a strong reuse of technologies mastered in previous versions of the 4G standard, both at the level of terminals and of the operators' infrastructure network [17].

1.4.6 LTE-M

For its part, LTE-M entered the race with impressive arguments compared to NB-IoT. Coexisting with 2G, 3G, and LTE, its technology does not require a change of base stations or infrastructure in general. It is nevertheless necessary to carry out a remote software update which involves unavailability of the network and a negligible cost. In addition, its higher transfer rates of 384 kpbs can support voice applications, its very low latency of around 50 ms allowing objects to communicate in motion such as cars and its communications based on TCP-IP protocol have attracted a growing number of players [17].

1.5 CHALLENGES AND ISSUES OF LPWAN

LPWAN technologies are proposed and developed in the licensed and unlicensed standard. Here are some challenges and issues faced by the deployment of LPWAN technologies [18].

1.5.1 Standardization

LPWANs are a new type of network for the IoT, with more than ten solutions offered. Even if some technologies will prevail, the need for standardization is important. Thus, in 2014, European Telecommunications Standards Institute (ETSI) proposed the Low Throughput Network (LTN) standard for low-speed LPWAN solutions. A functional architecture as well as interfaces and protocols are defined without imposing the Physical layer. Sigfox and LoRa are involved in this standardization. There is no single sub-1GHz ISM band in the world. For example, in Europe, we have the 868 MHz band, while in the United States, it is the 915 MHz band. The 2.4 GHz band is common in the world, but it is less interesting in terms of range [19].

1.5.2 Security

Downstream from the gateway, the equipment encountered implements conventional encryption and authentication mechanisms. These materials are not constrained in terms of resources. On the radio part, we are dealing with wireless communications, which are admittedly often encrypted and authenticated but which are ubiquitous by nature. Security management is, therefore, complex since the signals transported can be easily captured by attackers equipped with spy circuits dedicated to a protocol or generic Software-Defined Radio (SDR). The synthesis of signals by attackers is also possible with this type of device. The challenge is in the implementation of defence and prevention systems in objects that are low on resources. It should be noted that a technology like Sigfox does not provide by default the encryption of radio communications, it is up to the developer to manage this aspect if he wishes. The Over The Air update of the object's firmware is complex or even impossible in the case of low-speed LPWAN networks. This is a problem in the sense that, during the life of an object, vulnerabilities can appear without the possibility of updating its firmware [20].

1.5.3 NB-IoT and LTE-M and their integration into 5G

The arrival of cellular networks for IoT namely NB-IoT and LTE-M, already deployed in France in 2020 and which will be an integral part of 5G, is likely to significantly redesign the LPWAN landscape in the years to come. Greater speed, existence of a Quality of Service, low latency, and roaming are important advantages to be counterbalanced, however, with higher power consumption and operating costs than those of LPWAN networks such as Sigfox or LoRaWAN [21].

As for these latter networks, it is precisely in view of their very low consumption and very low-cost aspects that they will be able to continue their development by completing the cellular networks for the IoT. In addition, the possibility with LoRaWAN of being able to set up a private network without going through an operator is an advantage over its competitors [22].

1.6 CONCLUSION

The advancement of the Internet of Things, which aims to give any object the ability to connect to the Internet and communicate with other objects, has led to the emergence of new wireless communication technologies. In this context, we have exploited the performance and characteristics of wide coverage and low-power LPWAN networks. We first presented the LPWAN networks and its main fields of application and then we defined the most popular LPWAN technologies. This chapter presents a technical comparison of different LPWAN standards in terms of their main characteristics, each of them presents its own advantages and disadvantages depending on the type of applications considered. To meet LPWA design goals, this chapter has shown that there are still issues which need to be clarified despite the evolution of technologies.

REFERENCES

[1] Raza, Usman, Parag Kulkarni, and Mahesh Sooriyabandara. "Low power wide area networks: An overview." *IEEE Communications Surveys & Tutorials* 19.2 (2017): 855–873.

[2] Chaudhari, Bharat S., and Marco Zennaro, eds. *LPWAN Technologies for IoT and M2M Applications*. Academic Press, Elsevier, USA, 2020.

[3] Jradi, Hassan, et al. "Overview of the mobility related security challenges in LPWANs." *Computer Networks* 186 (2020): 107761.

[4] Al-Kashoash, Hayder A. A., and Andrew H. Kemp. "Comparison of 6LoWPAN and LPWAN for the Internet of Things." *Australian Journal of Electrical and Electronics Engineering* 13.4 (2016): 268–274.

[5] Adefemi Alimi, Kuburat Oyeranti, et al. "A survey on the security of low power wide area networks: Threats, challenges, and potential solutions." *Sensors* 20.20 (2020): 5800.

[6] Muteba, Franck, Karim Djouani, and Thomas Olwal. "A comparative survey study on LPWA IoT technologies: Design, considerations, challenges and solutions." *Procedia Computer Science* 155 (2019): 636–641.

[7] Pham, Tung Lam, et al. "Low power wide area network technologies for smart cities applications." *2019 International Conference on Information and Communication Technology Convergence (ICTC)*. IEEE, Jeju Island, Korea, 2019.

[8] Bembe, Mncedisi, et al. "A survey on low-power wide area networks for IoT applications." *Telecommunication Systems* 71.2 (2019): 249–274.

[9] Chaudhari, Bharat S., Marco Zennaro, and Suresh Borkar. "LPWAN technologies: Emerging application characteristics, requirements, and design considerations." *Future Internet* 12.3 (2020): 46.

[10] Rubio-Aparicio, Jesus, et al. "Design and implementation of a mixed IoT LPWAN network architecture." *Sensors* 19.3 (2019): 675.

[11] Thoen, Bart, et al. "A deployable LPWAN platform for low-cost and energy-constrained IoT applications." *Sensors* 19.3 (2019): 585.

[12] Centenaro, Marco, et al. "Long-range communications in unlicensed bands: The rising stars in the IoT and smart city scenarios." *IEEE Wireless Communications* 23.5 (2016): 60–67.

[13] Hwang, Seung-Hoon and Shu-Zhi Liu. "Survey on 3GPP low power wide area technologies and its application." *2019 IEEE VTS Asia Pacific Wireless Communications Symposium (APWCS)*. IEEE, Singapore, 2019.

[14] Lavric, Alexandru, Adrian I. Petrariu, and Valentin Popa. "Sigfox communication protocol: The new era of IoT?." *2019 International Conference on Sensing and Instrumentation in IoT Era (ISSI)*. IEEE, 2019.

[15] Croce, Daniele, et al. "Impact of LoRa imperfect orthogonality: Analysis of link-level performance." *IEEE Communications Letters* 22.4 (2018): 796–799.

[16] Queralta, J. Peña, et al. "Comparative study of LPWAN technologies on unlicensed bands for M2M communication in the IoT: Beyond LoRa and LoRaWAN." *Procedia Computer Science* 155 (2019): 343–350.

[17] Dawaliby, Samir, Abbas Bradai, and Yannis Pousset. "In depth performance evaluation of LTE-M for M2M communications." *2016 IEEE 12th International Conference on Wireless and Mobile Computing, Networking and Communications (WiMob)*. IEEE, New York, USA, 2016.

[18] Mekki, Kais, et al. "A comparative study of LPWAN technologies for large-scale IoT deployment." *ICT Express* 5.1 (2019): 1–7.

[19] Thubert, Pascal, Alexander Pelov, and Suresh Krishnan. "Low-power wide-area networks at the IETF." *IEEE Communications Standards Magazine* 1.1 (2017): 76–79.

[20] Torres, Nuno, Pedro Pinto, and Sérgio Ivan Lopes. "Security Vulnerabilities in LPWANs – An attack vector analysis for the IoT ecosystem." *Applied Sciences* 11.7 (2021): 3176.

[21] Aldahdouh, Khaled Aldahdouh, Khalid A. Darabkh, and Waleed Al-Sit. "A survey of 5G emerging wireless technologies featuring LoRaWAN, Sigfox, NB-IoT and LTE-M." *2019 International Conference on Wireless Communications Signal Processing and Networking (WiSPNET)*. IEEE, Chennai, India, 2019.

[22] Mwakwata, Collins Burton, et al. "Narrowband Internet of Things (NB-IoT): From physical (PHY) and media access control (MAC) layers perspectives." *Sensors* 19.11 (2019): 2613.

Chapter 2

LPWAN communication-based ML for IoT networks

Lokaiah Pullagura
Jain University

Nalli. Vinaya Kumari
Malla Reddy Institute of Technology & Science

Siva Kumar Gowda Katta
Sreenidhi Institute of Science and Technology

2.1 INTRODUCTION

The combination of Machine Learning (ML) and Low Power Wide Area Networks (LPWAN) has led to significant advancements in the field of Internet of Things (IoT) applications [1]. ML refers to a subset of Artificial Intelligence that enables systems to learn from data, make predictions, and decisions based on patterns, and improve over time. LPWAN, on the other hand, provides long-range and low-power connectivity, making it ideal for IoT devices that are usually deployed in remote or hard-to-reach locations [2,3].

The use of LPWAN for IoT applications has grown in recent years due to its low cost and low power consumption, enabling the deployment of a large number of sensors over a wide area [4]. ML can further enhance the functionality of IoT applications by providing insights and predictions based on the data collected by sensors [5]. By analyzing large volumes of data, ML algorithms can detect patterns, identify anomalies, and predict future outcomes, which can help optimize processes, reduce costs, and improve decision-making [6].

LPWAN and ML can be applied in various industries and use cases, including smart cities, agriculture, environmental monitoring, healthcare, and logistics [7,8]. For example, in the healthcare industry, LPWAN-based IoT devices can be used to monitor patient health, while ML can analyze the data and detect any anomalies that require attention. In the logistics industry, LPWAN can be used to track shipments and monitor inventory, while ML can optimize supply chain operations, as shown in Figure 2.1.

Overall, the combination of ML and LPWAN-based IoT applications holds great promise for improving the efficiency and effectiveness of various

DOI: 10.1201/9781003426974-2

Figure 2.1 LPWAN applications.

industries, providing insights and predictions based on real-time data, and ultimately leading to better decision-making and outcomes [6,9].

2.2 OVERVIEW OF LPWAN-BASED IoT SENSORS AND NETWORKS

LPWAN technology has emerged as a popular solution for connecting IoT sensors and devices over long distances with low power consumption [10]. In this overview, we cover the key aspects of LPWAN-based IoT sensors and networks in detail [11].

2.2.1 What is LPWAN?

LPWAN is a wireless communication technology designed to connect low-power IoT devices over long distances [12]. LPWAN technologies provide long-range, low-power, and low-cost connectivity for IoT sensors and devices. Some of the most popular LPWAN technologies include LoRaWAN, Sigfox, and NB-IoT [13].

2.2.2 Types of LPWAN-based IoT sensors

LPWAN-based IoT sensors come in various types, depending on the data they collect and the application they are used for. Some common types include:

- **Temperature sensors:** Used to monitor temperature changes in environments like buildings, warehouses, and cold storage rooms [14].
- **Humidity sensors:** Measure humidity levels in the air and are used in applications such as greenhouse monitoring and HVAC systems [15].
- **Motion sensors:** Detect motion and are used in security systems, home automation, and energy management.
- **Pressure sensors:** Measure substance pressure and are used in industrial automation and oil and gas monitoring.
- **Light sensors:** Measure light levels in an environment and are used in smart lighting and building automation [16].

2.2.3 LPWAN-based IoT Network Architecture

LPWAN-based IoT networks typically have a three-tier architecture:

- **Sensor layer:** Consists of IoT sensors that collect data from the environment and transmit it to the network [17].
- **Gateway layer:** Includes gateways that receive data from the sensors and transmit it to the cloud or server.
- **Cloud or server layer:** Stores and processes the sensor data, providing services like data analytics, visualization, and management [18].

2.2.4 Advantages of LPWAN-based IoT sensors and networks

LPWAN-based IoT sensors and networks offer several advantages over traditional wireless networks, including:

- **Long-range connectivity:** Enables IoT devices to connect over long distances, even in remote locations [19].
- **Low power consumption:** LPWAN-based sensors and devices consume minimal power, allowing them to operate for months or even years on a single battery [20].
- **Cost-effectiveness:** LPWAN-based sensors and devices are relatively inexpensive compared to other IoT technologies [21].
- **Scalability:** LPWAN-based IoT networks can easily scale up or down based on application requirements [22].

2.2.5 Challenges and limitations of LPWAN-based IoT sensors and networks

Despite their advantages, LPWAN-based IoT sensors and networks face some challenges and limitations, such as:

- **Limited bandwidth:** LPWAN networks have restricted bandwidth, making them unsuitable for high-speed data transmission applications [23].
- **Security concerns:** LPWAN-based IoT networks are susceptible to security breaches, including hacking and data theft [24].
- **Interference:** LPWAN networks can experience interference from other wireless networks, potentially resulting in data loss or corruption [25].
- **Limited data capacity:** LPWAN-based IoT sensors and devices can transmit only a limited amount of data, making them unsuitable for high-volume data transmission applications [26].

In conclusion, LPWAN-based IoT sensors and networks offer a cost-effective and scalable solution for connecting low-power IoT devices over long distances. However, they also have limitations that need to be considered for specific applications [27,28].

2.3 MACHINE LEARNING TECHNIQUES FOR LPWAN-BASED IoT APPLICATIONS

ML techniques have become increasingly popular in LPWAN-based IoT applications to enable more advanced data analytics and decision-making. Here are some of the most commonly used ML techniques in LPWAN-based IoT applications:

2.3.1 Anomaly detection

Anomaly detection is a ML technique that involves identifying unusual patterns or events in data. In LPWAN-based IoT applications, anomaly detection can be used to detect abnormal sensor readings, such as sudden spikes in temperature or humidity levels [29]. Anomaly detection can trigger alerts or notifications to prevent equipment damage, improve quality control, or ensure the safety of people and assets [30,31].

2.3.2 Predictive maintenance

Predictive maintenance is a ML technique that involves using data analytics to predict equipment failures or maintenance needs. In LPWAN-based IoT applications, predictive maintenance can detect early signs of equipment wear and tear and schedule maintenance activities before equipment failure occurs [32,33]. Predictive maintenance helps reduce maintenance costs, minimize equipment downtime, and improve equipment reliability [34].

2.3.3 Classification

Classification is a ML technique that involves identifying patterns or categories in data. In LPWAN-based IoT applications, classification can identify different types of sensor readings, such as temperature or humidity levels, and group them into different categories [35]. Classification improves quality control, optimizes energy consumption, and enhances asset management [36,37].

2.3.4 Regression

Regression is a ML technique that involves predicting a continuous value based on input data. In LPWAN-based IoT applications, regression can predict future sensor readings, such as temperature or humidity levels, based on past readings [38,39]. Regression optimizes energy consumption, improves quality control, and forecasts equipment performance [40].

2.3.5 Clustering

Clustering is a ML technique that involves grouping similar data points together based on their characteristics. In LPWAN-based IoT applications, clustering can group similar sensors or devices together based on their performance or location [41,42]. Clustering optimizes energy consumption, improves asset management, and facilitates predictive maintenance.

ML techniques have become essential tools in LPWAN-based IoT applications to enable more advanced data analytics and decision-making. Anomaly detection, predictive maintenance, classification, regression, and clustering are just a few examples of the many ML techniques that can be used in LPWAN-based IoT applications [43].

2.4 CHALLENGES AND LIMITATIONS OF MACHINE LEARNING AND LPWAN-BASED IoT APPLICATIONS

While ML has the potential to bring significant benefits to LPWAN-based IoT applications, there are also a number of challenges and limitations that should be considered:

2.4.1 Data quality and quantity

One of the biggest challenges of ML in LPWAN-based IoT applications is ensuring that the data is of high quality and quantity [44,45]. The accuracy and reliability of ML models depend on the quality and quantity of the data used to train them [46]. In LPWAN-based IoT applications, data can be affected by noise, interference, and other factors that can reduce its quality [29]. Additionally, LPWAN-based networks are designed to be low-power and low-bandwidth, which means that they may not generate enough data to train sophisticated ML models [47].

2.4.2 Data security and privacy

Another challenge of ML in LPWAN-based IoT applications is ensuring data security and privacy [48]. LPWAN-based networks can transmit data over long distances, making it vulnerable to interception and hacking [49]. Additionally, the data generated by LPWAN-based IoT sensors can be highly sensitive, making it important to protect it from unauthorized access [50].

2.4.3 Model complexity

ML models can become very complex, which can make them difficult to understand and interpret. This can be especially problematic in LPWAN-based IoT applications, where the models need to be small enough to run on low-power devices with limited processing capabilities [12].

2.4.4 Power consumption

ML algorithms can be computationally expensive, which can be problematic in LPWAN-based IoT applications where the devices have limited power resources. Running complex ML models can quickly drain the battery of an LPWAN-based IoT sensor, leading to reduced device lifetimes and increased maintenance costs [47].

2.4.5 Cost and deployment complexity

Finally, ML can be expensive to implement and maintain. It may require specialized expertise to develop, train, and deploy ML models, which can be a challenge for small and medium-sized enterprises. Additionally, deploying LPWAN-based IoT sensors and networks can require significant infrastructure investments, acting as a barrier to entry for some organizations [51–53].

In summary, while ML has the potential to bring significant benefits to LPWAN-based IoT applications, there are also a number of challenges and limitations that should be considered. These include data quality and quantity, data security and privacy, model complexity, power consumption, and cost and deployment complexity [54,55].

2.5 FUTURE DIRECTIONS FOR MACHINE LEARNING AND LPWAN-BASED IoT APPLICATIONS

ML and LPWAN-based IoT applications are rapidly evolving, and several future directions are worth considering. Here are some potential areas for development:

2.5.1 Edge computing and ML

One promising future direction for ML in LPWAN-based IoT applications is the integration of edge computing [56]. Edge computing involves moving the data processing closer to the sensors, which can help reduce the amount of data that needs to be transmitted over the network [57]. This can help reduce the power consumption of the sensors and improve the scalability of the network. ML algorithms can be integrated with edge computing to enable real-time decision-making and reduce the amount of data that needs to be transmitted to the cloud [58,59].

2.5.2 Federated learning

Federated learning is a distributed ML technique that involves training models on data that is stored on multiple devices [30]. This can be particularly useful in LPWAN-based IoT applications, where the devices are distributed and have limited power resources [26]. Federated learning can enable more efficient and scalable ML models, and it can also help address privacy concerns by keeping the data local.

2.5.3 Hybrid models

Hybrid ML models that combine deep learning and classical ML techniques can be particularly useful in LPWAN-based IoT applications. Deep learning can be used to extract meaningful features from the data, while classical ML techniques can be used to make predictions and decisions [56,59]. This can enable more accurate and reliable ML models that can be trained with smaller datasets [60].

2.5.4 Standardization

Standardization is essential for the interoperability and scalability of LPWAN-based IoT applications. The development of common data models, interfaces, and protocols can help ensure that different LPWAN-based IoT devices and networks can work together seamlessly [61,62]. Standardization can also help reduce the development and deployment costs of LPWAN-based IoT applications [63].

2.5.5 Low-power ML algorithms

Finally, the development of low-power ML algorithms that can run on low-power devices is critical for the widespread adoption of LPWAN-based IoT applications [64]. The development of new ML algorithms that can run on low-power devices, such as microcontrollers and sensor nodes, can enable more sophisticated decision-making capabilities without significantly increasing the power consumption of the devices [65,66].

In summary, there are several future directions for ML and LPWAN-based IoT applications, including the integration of edge computing and ML, federated learning, hybrid models, standardization, and low-power ML algorithms. These developments can help improve the scalability, reliability, and efficiency of LPWAN-based IoT applications and enable their widespread adoption in a range of industries [67].

2.6 CONCLUSION

In conclusion, ML and LPWAN-based IoT applications have the potential to bring significant benefits to a wide range of industries. ML can be used to extract insights and make predictions from large and complex datasets, while LPWAN-based IoT networks can enable the widespread deployment of low-power and low-cost sensors and devices. The combination of ML and LPWAN-based IoT can enable more accurate and reliable decision-making capabilities in various fields, such as smart cities, agriculture, healthcare, and industry.

However, there are also several challenges and limitations that need to be addressed, such as data quality and quantity, data security and privacy, model complexity, power consumption, and cost and deployment complexity. To overcome these challenges, future developments in edge computing and ML, federated learning, hybrid models, standardization, and low-power ML algorithms will be essential.

Overall, ML and LPWAN-based IoT applications are rapidly evolving, and there is significant potential for continued innovation and development in the future. As these technologies become more widely adopted and integrated, they have the potential to transform industries and enable new applications and use cases that were previously not possible.

REFERENCES

[1] Raza, U.; Kulkarni, P.; Sooriyabandara, M. Low Power Wide Area Networks: An Overview. *IEEE Commun. Surv. Tutor.* 2017, 19, 855–873.

[2] Gartner Identifies the Top 10 Internet of Things Technologies for 2017 and 2018. Available online: https://www.gartner.com/en/newsroom/press-releases/2016-02-23-gartner-identifies-the-top-10-internet-of-things-technologies-for-2017-and-2018 (accessed on 14 May 2021).

[3] Chaudhari, B.S.; Zennaro, M.; Borkar, S. LPWAN Technologies: Emerging Application Characteristics, Requirements, and Design Considerations. *Future Internet* 2020, 12, 46.

[4] Melakessou, F.; Kugener, P.; Alnaffakh, N.; Faye, S.; Khadraoui, D. Heterogeneous Sensing Data Analysis for Commercial Waste Collection. *Sensors* 2020, 20, 978.

[5] Abo-Tabik, M.; Costen, N.; Darby, J.; Benn, Y. Towards a Smart Smoking Cessation App: A 1D-CNN Model Predicting Smoking Events. *Sensors* 2020, 20, 1099.

[6] Pradhan, B.; Bhattacharyya, S.; Pal, K. IoT-Based Applications in Healthcare Devices. *J. Healthc. Eng.* 2021, 2021, 663259

[7] Sun, J.; Fang, Y.; Zhu, X. Privacy and Emergency Response in e-Healthcare Leveraging Wireless Body Sensor Networks. *IEEE Wirel. Commun.* 2010, 17, 66–73.

[8] Liu, T.; Xu, H.; Ragulskis, M.; Cao, M.; Ostachowicz, W. A Data-Driven Damage Identification Framework Based on Transmissibility Function Datasets and One-Dimensional Convolutional Neural Networks: Verification on a Structural Health Monitoring Benchmark Structure. *Sensors* 2020, 20, 1059.

[9] She, J.; Jiang, J.; Ye, L.; Hu, L.; Bai, C.; Song, Y. 2019 Novel Coronavirus of Pneumonia in Wuhan, China: Emerging Attack and Management Strategies. *Clin. Transl. Med.* 2020, 9, 1–7.

[10] Preventing the Spread of the Coronavirus. Available online: https://www.health.harvard.edu/diseases-and-conditions/preventing-the-spread-of-the-coronavirus (accessed on 10 June 2021).

[11] Ullah, A.; Azeem, M.; Ashraf, H.; Alaboudi, A.A.; Humayun, M.; Jhanjhi, N. Secure Healthcare Data Aggregation and Transmission in IoT-A Survey. *IEEE Access* 2021, 9, 16849–16865.

[12] Olatinwo, D.D.; Abu-Mahfouz, A.; Hancke, G. A survey on LPWAN Technologies in WBAN for Remote Health-Care Monitoring. *Sensors* 2019, 19, 5268.

[13] Qadri, Y.A.; Nauman, A.; Zikria, Y.B.; Vasilakos, A.V.; Kim, S.W. The Future of Healthcare Internet of Things: A Survey of Emerging Technologies. *IEEE Commun. Surv. Tutor.* 2020, 22, 1121–1167.

[14] Khan, Z.A.; Abbasi, U. Evolution of Wireless Sensor Networks toward Internet of Things. In *Emerging Communication Technologies Based on Wireless Sensor Networks: Current Research and Future Applications*; CRC Press: Boca Raton, FL, 2016; Chapter 7; pp. 179–200.

[15] National Intelligence Council, Disruptive Civil Technologies-Six Technologies with Potential Impacts on US Interests Out to 2025. Available online: https://globaltrends.thedialogue.org/publication/disruptive-civil-technologies-six-technologies-with-potential-impacts-on-us-interests-out-to-2025/ (accessed on 17 September 2020).

[16] Giusto, D.; Iera, A.; Morabito, G.; Atzori, L. *The Internet of Things: 20th Tyrrhenian Workshop on Digital Communications*; Springer Science & Business Media: Berlin/Heidelberg, Germany, 2010.

[17] European Commission Report on Internet of Things. Available online: https://ec.europa.eu/digital-single-market/en/internet-of-things (accessed on 17 September 2020).

[18] International Telecomunicaciones Union (ITU). Recommendation ITU-T Y.4000/Y.2060 (2012), Overview of the Internet of Things. 2012. Available online: https://www.itu.int/rec/T-REC-Y.2060-201206-I (accessed on 17 June 2020).

[19] Vermesan, O.; Friess, P.; Guillemin, P.; Gusmeroli, S.; Sundmaeker, H.; Bassi, A.; Jubert, I.; Mazura, M.; Harrison, M.; Eisenhauer, M.; et al. Internet of Things Strategic Research Roadmap. 2009. Available online: https://www.sintef.no/en/publications/publication/?pubid=CRIStin+911003 (accessed on 17 September 2020).

[20] Ray, P. A Survey on Internet of Things Architectures. *J. King Saud Univ. Comput. Inf. Sci.* 2018, 30, 291–319.

[21] Ray, B. The Past, Present, and Future of LPWAN. Available online: https://www.link-labs.com/blog/past-present-future-lpwan (accessed on 10 June 2020).

[22] Low-Power Wide Area Network (LPWAN) Overview. Available online: https://tools.ietf.org/pdf/rfc8376.pdf (accessed on 17 May 2020).

[23] European Commission. MHealth, What Is It?-Infographic. Available online: https://ec.europa.eu/digital-single-market/en/news/mhealth-what-it-info-graphic (accessed on 17 May 2020).

[24] Global Asthma Network: The Global Asthma Report for 2014. Available online: https://www.globalasthmareport.org/ (accessed on 19 May 2020).

[25] Foubert, B.; Mitton, N. Long-Range Wireless Radio Technologies: A Survey. *Future Internet* 2020, 12, 13.

[26] Roland, D.; Lantauro, I. Google Parent and Sanofi Name Diabetes Joint Venture Onduo. Available online: https://www.wsj.com/articles/google-parent-and-sanofi-name-diabetes-joint-venture-onduo-1473659627 (accessed on 19 May 2020).

[27] Amadou, I.; Foubert, B.; Mitton, N. LoRa in a Haystack: A study of the LORA Signal Behavior. In Proceedings of the 2019 International Conference on Wireless and Mobile Computing, Networking and Communications (WiMob), Barcelona, Spain, 21–23 October 2019; pp. 1–4

[28] Nugraha, A.T.; Hayati, N.; Suryanegara, M. The Experimental Trial of LoRa System for Tracking and Monitoring Patient with Mental Disorder. In Proceedings of the 2018 International Conference on Signals and Systems (ICSigSys), Bali, Indonesia, 1–3 May 2018; pp. 191–196.

[29] Kim, T.; Vecchietti, L.F.; Choi, K.; Lee, S.; Har, D. Machine Learning for Advanced Wireless Sensor Networks: A Review. *IEEE Sens. J.* 2021, 21, 12379–12397.

[30] Gu, T. Detection of Small Floating Targets on the Sea Surface Based on Multi-Features and Principal Component Analysis. *IEEE Geosci. Remote Sens. Lett.* 2020, 17, 809–813.

[31] Li, Y.F.; Guo, L.Z.; Zhou, Z.H. Towards Safe Weakly Supervised Learning. *IEEE Trans. Pattern Anal. Mach. Intell.* 2021, 43, 334–346.

[32] Wang, X.; Kihara, D.; Luo, J.; Qi, G.J. EnAET: A Self-Trained Framework for Semi-Supervised and Supervised Learning with Ensemble Transformations. *IEEE Trans. Image Process.* 2021, 30, 1639–1647.

[33] Shang, C.; Chang, C.Y.; Chen, G.; Zhao, S.; Lin, J. Implicit Irregularity Detection Using Unsupervised Learning on Daily Behaviors. *IEEE J. Biomed. Health Inform.* 2020, 24, 131–143.

[34] Hoang, T.M.; Duong, T.Q.; Tuan, H.D.; Lambotharan, S.; Hanzo, L. Physical Layer Security: Detection of Active Eavesdropping Attacks by Support Vector Machines. *IEEE Access* 2021, 9, 31595–31607.

[35] Daza, E.J.; Wac, K.; Oppezzo, M. Effects of Sleep Deprivation on Blood Glucose, Food Cravings, and Affect in a Non-Diabetic: An N-of-1 Randomized Pilot Study. *Healthcare* 2020, 8, 6.

[36] European Commission. Green Paper on Mobile Health. Available online: https://ec.europa.eu/digital-single-market/en/news/green-paper-mobile-health-mhealth (accessed on 17 May 2020).

[37] European Commission. Public Consultation on the Green Paper on Mobile Health. Available online: https://digital-strategy.ec.europa.eu/en/library/summary-report-public-consultation-green-paper-mobile-health (accessed on 17 May 2020).

[38] 5G and e-Health. Available online: https://5g-ppp.eu/wp-content/uploads/2016/02/5G-PPP-White-Paper-on-eHealth-Vertical-Sector.pdf (accessed on 17 May 2020).

[39] Thuemmler, C.; Paulin, A.; Jell, T.; Lim, A.K. Information Technology-Next Generation: The Impact of 5G on the Evolution of Health and Care Services. In *Information Technology-New Generations*; Latifi, S., Ed.; Springer International Publishing: Cham, Swizerland, 2018; pp. 811–817.

[40] Mekki, K.; Bajic, E.; Chaxel, F.; Meyer, F. A Comparative Study of LPWAN Technologies for Large-Scale IoT Deployment. *ICT Express* 2019, 5, 1–7.

[41] Khan, Z.A.; Auguin, M. A Multichannel Design for QoS Aware Energy Efficient Clustering and Routing in WMSN. *Int. J. Sen. Netw.* 2013, 13, 145–161.

[42] An Introduction to Machine Learning. Available online: https://www.digitalocean.com/community/tutorials/an-introduction-to-machine-learning (accessed on 10 January 2020).

[43] Zhang, Z.; Wang, D.; Gao, J. Learning Automata-Based Multiagent Reinforcement Learning for Optimization of Cooperative Tasks. *IEEE Trans. Neural Netw. Learn. Syst.* 2020, 1–14.

[44] Zhang, W.; Chen, X.; Liu, Y.; Xi, Q. A Distributed Storage and Computation k-Nearest Neighbor Algorithm Based Cloud-Edge Computing for Cyber-Physical-Social Systems. *IEEE Access* 2020, 8, 50118–50130.

[45] Bharadwaj, H.K.; Agarwal, A.; Chamola, V.; Lakkaniga, N.R.; Hassija, V.; Guizani, M.; Sikdar, B. A Review on the Role of Machine Learning in Enabling IoT Based Healthcare Applications. *IEEE Access* 2021, 9, 38859–38890.

[46] Posonia, A.M.; Vigneshwari, S.; Rani, D.J. Machine Learning based Diabetes Prediction Using Decision Tree J48. In Proceedings of the 2020 3rd International Conference on Intelligent Sustainable Systems (ICISS), Tirunelveli, India, 17–18 December 2020; pp. 498–502.

[47] Kaur, R.; Singh, A.; Singla, J. Integration of NIC algorithms and ANN: A Review of Different Approaches. In Proceedings of the 2021 2nd International Conference on Computation, Automation and Knowledge Management (ICCAKM), Dubai, United Arab Emirates, 19–21 January 2021; pp. 185–190.

[48] Istepanian, R.S.H.; Hu, S.; Philip, N.Y.; Sungoor, A. The Potential of Internet of m-Health Things (m-IoT) for Non-Invasive Glucose Level Sensing. In Proceedings of the 2011 Annual International Conference of the IEEE Engineering in Medicine and Biology Society, Boston, MA, 30 August–3 September 2011; pp. 5264–5266.

[49] Lijun, Z. Multi-Parameter Medical Acquisition Detector Based on Internet of Things. *Chin. Pat.* 2013, 202, 774.

[50] Niu, L.; Chen, C.; Liu, H.; Zhou, S.; Shu, M. A Deep-Learning Approach to ECG Classification Based on Adversarial Domain Adaptation. *Healthcare* 2020, 8, 437.

[51] Jara, A.J.; Zamora-Izquierdo, M.A.; Skarmeta, A.F. Interconnection Framework for mHealth and Remote Monitoring Based on the Internet of Things. *IEEE J. Sel. Areas Commun.* 2013, 31, 47–65.

[52] Rasid, M.F.A.; Musa, W.M.W.; Kadir, N.A.A.; Noor, A.M.; Touati, F.; Mehmood, W.; Khriji, L.; Al-Busaidi, A.; Ben Mnaouer, A. Embedded Gateway Services for Internet of Things Applications in Ubiquitous Healthcare. In Proceedings of the 2014 2nd International Conference on Information and Communication Technology (ICoICT), Bandung, Indonesia, 28–30 May 2014; pp. 145–148.

[53] Athaya, T.; Choi, S. An Estimation Method of Continuous Non-Invasive Arterial Blood Pressure Waveform Using Photoplethysmography: A U-Net Architecture-Based Approach. *Sensors* 2021, 21, 1867.

[54] Dash, P. Electrocardiogram Monitoring. *Indian J. Anaesth.* 2002, 46, 251–260.

[55] Yang, G.; Xie, L.; Mantysalo, M.; Zhou, X.; Pang, Z.; Xu, L.D.; Kao-Walter, S.; Chen, Q.; Zheng, L. A Health-IoT Platform Based on the Integration of Intelligent Packaging, Unobtrusive Bio-Sensor, and Intelligent Medicine Box. *IEEE Trans. Ind. Inform.* 2014, 10, 2180–2191.

[56] Hoang, M.L.; Carratù, M.; Paciello, V.; Pietrosanto, A. Body Temperature-Indoor Condition Monitor and Activity Recognition by MEMS Accelerometer Based on IoT-Alert System for People in Quarantine Due to COVID-19. *Sensors* 2021, 21, 2313.

[57] Ruiz, M.; García, J.; Fernández, B. Body Temperature and Its Importance as a Vital Constant. *Rev. Enferm.* 2009, 32, 44–52.

[58] Dohr, A.; Modre-Opsrian, R.; Drobics, M.; Hayn, D.; Schreier, G. The Internet of Things for Ambient Assisted Living. In Proceedings of the 2010 Seventh International Conference on Information Technology: New Generations, Las Vegas, NV, 12–14 April 2010; pp. 804–809.

[59] Puustjarvi, J.; Puustjarvi, L. Automating Remote Monitoring and Information Therapy: An Opportunity to Practice Telemedicine in Developing Countries. In Proceedings of the 2011 IST-Africa Conference Proceedings, Gaborone, Botswana, 11–13 May 2011; pp. 1–9.

[60] Jian, Z.; Zhanli, W.; Zhuang, M. Temperature Measurement System and Method Based on Home Gateway. *Chin. Pat.* 2012, 102, 185.

[61] In, Z.L. Patient Body Temperature Monitoring System and Device Based on Internet of Things. *Chin. Pat.* 2014, 103, 577–688.

[62] Zhang, Q.; Arney, D.; Goldman, J.M.; Isselbacher, E.M.; Armoundas, A.A. Design Implementation and Evaluation of a Mobile Continuous Blood Oxygen Saturation Monitoring System. *Sensors* 2020, 20, 6581.

[63] Khattak, H.A.; Ruta, M.; Eugenio Di Sciascio, E. CoAP-Based Healthcare Sensor Networks: A Survey. In Proceedings of the 2014 11th International Bhurban Conference on Applied Sciences Technology (IBCAST), Islamabad, Pakistan, 14–18 January 2014; pp. 499–503.

[64] Larson, E.C.; Goel, M.; Boriello, G.; Heltshe, S.; Rosenfeld, M.; Patel, S.N. SpiroSmart: Using a Microphone to Measure Lung Function on a Mobile Phone. In Proceedings of the 2012 ACM Conference on Ubiquitous Computing, Pittsburgh, PA, 5–8 September 2012; pp. 280–289.

[65] Larson, E.C.; Goel, M.; Redfield, M.; Boriello, G.; Rosenfeld, M.; Patel, S.N. Tracking Lung Function on any Phone. In Proceedings of the 3rd ACM Symposium on Computing for Development, Bangalore, India, 11–12 January 2013; pp. 1–2.

[66] Livingstone, R.; Paleg, G. Enhancing Function, Fun and Participation with Assistive Devices, Adaptive Positioning, and Augmented Mobility for Young Children with Infantile-Onset Spinal Muscular Atrophy: A Scoping Review and Illustrative Case Report. *Disabilities* 2021, 1, 1–22.

[67] Drew, B.J.; Califf, R.M.; Funk, M.; Kaufman, E.S.; Krucoff, M.W.; Laks, M.M.; Macfarlane, P.W.; Sommargren, C.; Swiryn, S.; Hare, G.F.V. Practice Standards for Electrocardiographic Monitoring in Hospital Settings. *Circulation* 2004, 110, 2721–2746.

Machine learning and LPWAN-based internet of things applications

Najia Ait Hammou and Hajar Mousannif
Cadi Ayyad University

Brahim Lakssir
Smart Devices Department MAScIR

3.1 INTRODUCTION

Robotics and the Internet of Things (IoT) are two distinct industries. A single technical niche has now been created nonetheless as a result of their merger. The phrase "the Internet of Robotic Things" refers to how IoT and robotics interact and function together. Robotics' development and adoption have made it possible for experts to create Internet of Things (IoT) applications based on robotics, creating a real-world possibility for the future. IoT and robots work together fundamentally to solve issues, develop new methods, support human endeavors, and carry out specific jobs. Automating this process helps to eliminate repetitive work and minimize human error. It is possible to simultaneously aim for better standards of progress as a result [1]. So, what exactly is robotics, and what does the IoT imply?

Robotics is a field that combines engineering and computer science. It covers robot design, manufacture, use, and operation. Robotics aims to create machines that can help and support people. It combines many academic fields, such as mechatronics, electronics, computer and software engineering, and others [2]. The term "Internet of Things" refers to a collection of electronic and physical items that have sensors and actuators to collect data about their surroundings. The devices are all linked together through Ethernet, Bluetooth, and Wi-Fi on the Internet. Following processing and analysis, notifications or alerts are then sent to end users on their individual devices based on the collected data [1].

At the moment, IoT applications incorporate a wide range of technologies. One of the most used technologies that meets the needs of specific areas is cellular networks (2G, 3G, 4G, and 5G), followed by Bluetooth, Wi-Fi, Zigbee, and others. New LPWAN technologies, however, provide long-distance communication with low power consumption and a slow data transmission rate. The three qualities mentioned above are unique to

DOI: 10.1201/9781003426974-3

agricultural applications; hence, the new LPWAN technologies are appropriate for the demands of this industry [3].

Applications have grown more intelligent as a result of the IoT revolution. Machine learning (ML) approaches are being used to significantly improve the intelligence and capabilities of an application as the number of data points collected rises [4].

With the help of the IoT, new technologies are being developed to address issues, improve processes, and boost productivity. The IoT is defined as the interconnection of uniquely identified electronic items over the Internet using "data plumbing" such as Internet Protocol (IP), cloud computing, and web services. It has a significant impact on industrial automation. His main objective is to facilitate communication and interaction between input and output from the production area, including analyzers, actuators, and robotics, to increase manufacturing and flexibility. Industrial automation has made significant commercial uses of technologies possible through the IoT [5]. On the other hand, IoT applications in agriculture have the power to change both the course of human and global history. The term "smart agriculture" refers to a wide range of agricultural and food production methods supported by the IoT and big data. When we discuss the IoT, we frequently mean how sensing, automation, and analytics technologies have been integrated into contemporary agricultural processes. So far, its most popular applications in smart agriculture are sensor-based systems and smart farm vehicles [6]. IoT smart farming products are made to assist farmers in monitoring crop fields using sensors and automating irrigation systems. The term "Smart Agri Robots" is used to describe weed-control robots [7].

First and foremost, this chapter describes the role of ML and LPWAN in IoT applications. Then it presents the four components of IoT architecture: (1) sensing, (2) network, (3) platform, and (4) application, before defining the architecture of agricultural IoT. Also, it provides an overview of the impact of IoT in agriculture and discusses the results of IoT applications in weeding.

3.2 RELATED WORKS

3.2.1 Machine learning

Automating a variety of jobs is possible with ML. The IoT is crucial for smart applications and offers services using efficient tech-based life-saving techniques.

The IoT is one of the most important technological advancements of the 21st century. It is a network of physically connected objects that can exchange data using sensors, software, and other technologies. When compared to a system that is human-centered, the IoT aims to be more efficient and transmit critical information more rapidly [8].

ML enables the IoT to glean hidden insights from the vast amount of sensed data. The IoT won't be able to fulfill future demands from organizations, governments, and individual users without ML. The main objective of the IoT is to understand what is going on in our surroundings and enable automated decision-making through intelligent approaches [9].

3.2.2 LPWAN

One of the fundamental protocols for the installation of the IoT is LPWAN (Low Power Wide Area Network), also referred to as LPWA or LPN. LPWAN is frequently mentioned as a reliable transport protocol choice for specific IoT requirements. In terms of technology, LPWAN satisfies the criteria of the IoT, thanks to its wide geographic coverage, high data transmission capacity, and low power consumption. Furthermore, it separated LPWAN from other wireless technologies such as Wi-Fi, Bluetooth, 3GPP, and Zigbee [10]. LPWAN systems make use of extremely constrained radio networks with little data transmission demand. As shown in Figure 3.1, the primary categories of LPWAN technologies include non-3GPP (LoRa, Sigfox, and Weightless) and 3GPP (NB-IoT, LTE-M, and EC-GSM). The most widely used technologies right now are Sigfox, NB-IoT, and LoRaWAN [11]. There are two categories of LPWAN technology: those that use licensed frequency bands and those that use unlicensed frequencies. Because LPWAN technologies enable long-range connections, low power consumption, and nearly real-time communication, they are the perfect IoT networking solution for IoT applications and devices [12]. Given the approaching universality of connected devices in many contexts, the family of technologies known as LPWAN is getting a lot of attention. The cellular architecture that its systems use allows for high node density and scalability. LPWAN solutions employ incredibly dependable modulation

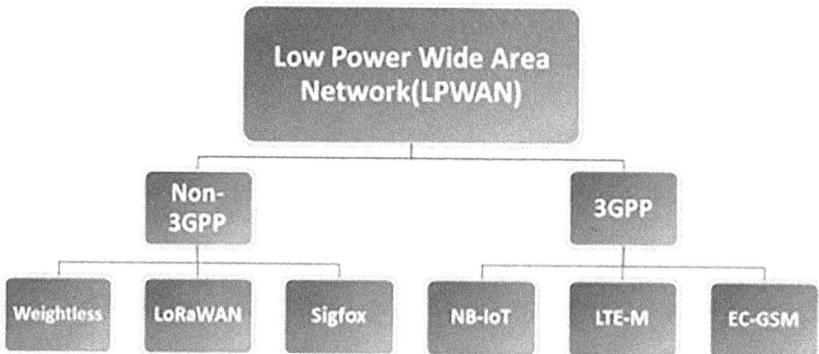

Figure 3.1 LPWAN classification.

and coding techniques. They make it possible to transmit data very efficiently over very long distances [13].

3.2.3 What is the IoT?

IoT has developed into one of the most significant 21st century technologies over the last few years, so we can use embedded devices to connect common objects to the Internet. Even if the digital and physical worlds collide, they work together. Physical things can share and gather data with little assistance from humans, thanks to low-cost computers and mobile technologies [14].

The IoT isn't just about connected gadgets; it's about the data those devices gather and the insightful conclusions drawn from it. Some examples of IoT applications include connected products, predictive maintenance, facility management, and remote monitoring [15].

3.2.4 Components of IoT architecture

An IoT architecture mainly consists of the following components, as shown in Figure 3.2.

- **Sensing, embedded components:** In the perception layer, you can find sensors, actuators, and other IoT devices that can perceive, compute, and connect to other devices. It provides accurate and dependable statistics. By detecting physical changes in the environment, sensors

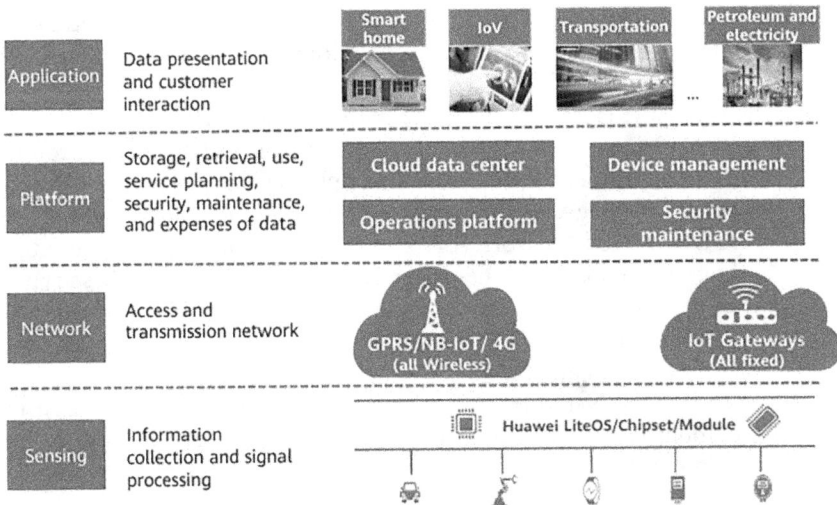

Figure 3.2 IoT architecture.

collect data. The IoT gateway, also known as the data collection layer, collects the data before it is integrated and processed [16].

- **Connectivity:** The pillars of networking, communication, and connectivity form the foundation of every IoT ecosystem. IoT protocols are used to send data between locations. The most widely used wireless protocols include Wi-Fi, Zigbee, LoRa, and cellular technologies. Data is transferred over gateways to servers or the cloud. Gateways provide security by preventing unwanted access [16].
- **Platform:** Every piece of incoming data is kept on the cloud, which is called the event processing layer. In this situation, the data is processed through data analysis, and decisions are made based on this data to trigger a system response. It is responsible for collecting all data and information coming from various IoT devices [16].
- **Application:** The application layer, often referred to as the API management layer, serves as the interface between third-party apps and the system or users. This is a legitimate process that maintains the data on hand and saves information for use in later responses [16].

3.2.5 Architecture of agricultural IoT system

It presents the application of the IoT architecture in the agricultural domain, as shown in Figure 3.3 [17].

- The object layer is a phase containing crops, livestock, and agricultural machinery.
- The perception layer contains geographic information systems (GIS), remote sensing systems (RS), and global navigation satellite systems (GNSS) that provide considerable opportunities to monitor and manage many facets of the systems for positioning, navigation, and timing [18].
- The transport layer includes three network types [19]:
  - PAN (Personal Area Network) is the smallest network type around a single person. It consists of smart phones, laptops, tablets, wearable technology, or any other personal digital device.
  - A Local Area Network (LAN) is a small type of network used in houses, companies, schools, or any other small area. These types of networks are used for file sharing, resource sharing, and communication in a small area. It is also used to share an Internet connection in any place.
  - Wide Area Network (WAN) is the largest network type used in computer networks. It covers large geographical areas, and it can also connect other small and medium networks like LANs and MANs.

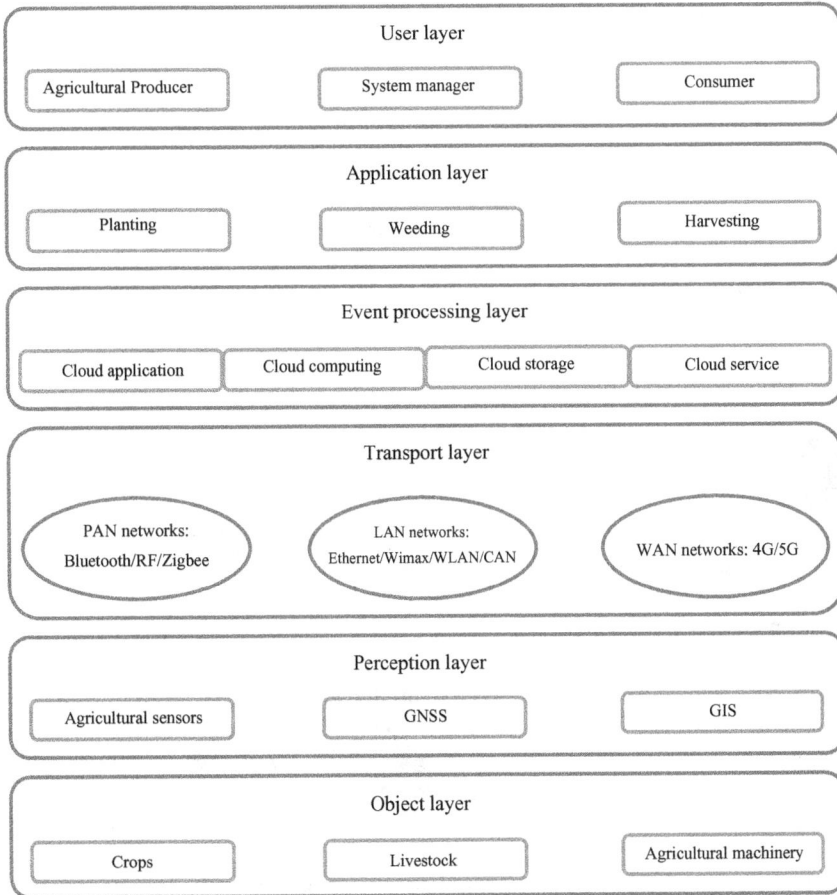

Figure 3.3 Architecture of agricultural IoT.

- The event processing layer covers:
  - Data may now be saved on the Internet in the same manner that it can be saved on a computer, thanks to a technique called cloud storage. On the other hand, specific projects are worked on and finished using cloud computing [20]. A cloud application is a piece of software that splits data storage and processing logic between client-side and server-side computing platforms [21]. Any service that is delivered from a cloud computing provider's server to users on demand through the Internet as opposed to a business's own on-premises servers is referred to as a "cloud service" [22].

- The application layer:
  - The act of weeding involves removing weeds from a field. It is a productive pre-harvest technique for managing crop yield and crop protection [23]. The act or occurrence of putting seeds or young plants into the soil is known as planting [24]. Harvesting is the act of gathering and cutting crops. Harvesting is the process of cutting a crop after it has reached maturity [25].
  - The user layer presents different users of this device as agricultural producers, system managers, and consumers that can benefit from it.

3.3 IoT APPLICATIONS

3.3.1 The impact of IoT in agriculture

The IoT is now a part of the fourth industrial revolution, known as Industry 4.0, where all factory parts are interconnected with the Internet to allow for real-time factory management and control. Agriculture 4.0 is the result of the industry 4.0 trend, which is being driven by the IoT and other information and communication technologies and is changing all industries' production capacities [26]. The IoT is replacing conventional farming methods. Then, innovation is crucial in order to create solutions that can solve urgent problems. The IoT offers a number of intelligent solutions that can assist in overcoming these difficulties. Smart agriculture has already had a substantial impact on the industry by maximizing resources, reducing waste, and raising farm productivity. With the use of digital technologies, farmers may use precision agriculture, crop monitoring, irrigation, and fertilizer management as methods of farm management to better decide when, where, and how much to fertilize, water, and spray pesticides on their crops [27].

3.3.2 LPWAN technology in agriculture

Both the cutting-edge private standard LoRa Symphony Link and the open network standard LoRaWAN serve as representations of LPWAN. Ingenu RPMA is known for its radically innovative technology, but Sigfox is known for its high availability and simplicity of usage. Compared to other technologies like Wi-Fi or Bluetooth, wireless LANs in agriculture have a definite benefit. Depending on the situation, these technologies can be employed successfully for agricultural purposes [28].

Lora is a new IoT communication technology used in many different industries that has significant development potential. To manage certain plant conditions in agriculture, environmental monitoring is necessary.

Bluetooth and low-power LPWAN are two communication methods that can be used in smart agricultural systems. Therefore, sensing devices can be utilized in the IoT to address issues affecting farmers [29]. Landowners can monitor the performance of their crops with the use of an IoT-based system. Finding a low-cost, low-power, long-range communication method is the major goal for the smart agriculture industry. An LPWAN-based sensor network topology can deliver real-time agricultural data to the end user [30].

3.3.3 ML in agriculture

Agriculture often provides for the basic needs of people and is seen as a key source of employment. Lack of understanding about market trends, soil types, yields, crops, and weather, as well as poor pesticide use, all contributed to the loss of farmers or increased costs. Therefore, monitoring data on crops aids farmers in making better decisions and resolving agricultural-related issues. The newest innovations in computer science are blockchain technology, cloud computing, the IoT, ML, and deep learning (DL). So, farmers and related industries will be able to produce more, improve quality, and ultimately make more money by using these technologies [31]. Smart agriculture includes the use of information and communication technologies (ICT). These technologies, which produce a lot of data, include farm management information systems, moisture and soil sensors, accelerometers, wireless sensor networks, cameras, drones, etc. So, ML has arisen to take advantage of the exponential growth in computing capability because traditional data processing methods cannot keep up with the ever-increasing needs of the new era of smart agriculture. When it comes to weed control, ML algorithms used in conjunction with imaging technology or non-imaging spectroscopy can distinguish and find specific weeds in real time. Thus, we can accurately access targeted locations to choose the shortest weed control path [32].

3.3.4 Application of IoT in weeding

Row crop cultivation makes up a considerable share of the world's total food production. One of the most crucial elements in the production of all agricultural products is weed management. Without adequate weed control, weeds will significantly reduce productivity by competing with agricultural plants for moisture, nutrients, and sunlight. For row crops, common weed management techniques include applying herbicides and tillage, as well as mechanical row harrowing [33]. Hence, the need to use IoT technology for the design of a weed control robot. The latter must be able to detect weeds and navigate between rows and intra-rows. An autonomous vehicle uses vision, wheel encoders, a pair of piezoresistive accelerometers, and a solid-state compass. The data from the sensors will be merged using a Kalman filter. To map the crop rows with the camera, computer vision will first recognize the positions of the plants. For guidance, a high-precision GPS is preferred. The energy will be drawn through solar panels to recharge the batteries.

Therefore, before considering the design of a weeding robot, we must look at its different movements. It will be easy to define its electronic system to determine which hardware should be integrated [34].

- For the placement of the weed control, dead reckoning is advised. This code wheel can be used to locate the actuator with respect to the planted area.
- When it comes to intra-row weeding, the crop will always suffer from non-targeted application, especially when considering the outcomes of electrical and chemical weeding techniques. As a result, the only method that can find and remove weeds is the use of mechanical actuators.
- GPS signals can be used to detect whether the robot is inside or outside the boundary. It is used to determine if the weed control robot is at the headland or within a field.
- Thanks to machine vision, the weeding robot can navigate the row without needing precise coordinates for a path to follow. Machine vision can instantly identify crop rows.
- Locomotion duties entail providing the robot with energy through an electrical connector that charges batteries attached to the robot. The robot has solar panels that may be used to get energy from the sun. Four wheels are better for a light vehicle since they provide better grip for in-row weeding.
- The on-board computer that was used to enter the robot's parameters is the most trustworthy method of communication with a nearby user. SMS messages are used to inform users when a GSM network is accessible. An easily navigable and well-organized web page offers an excellent opportunity to provide facts regarding the system's status.
- The super canopy's circumferential detection provides a fundamental level of safety. On the weeding robot's four sides, it detects obstructions above the crop plants.

3.4 CONCLUSION

The IoT is one of the technologies of the 21st century. It describes the network that integrates sensors and software to connect and exchange data. Its combination with robotics gives birth to the term "internet of robotic objects." Using ML, it is possible to progress from simple manual tasks to much more complex ones. The latter is indeed a category of artificial intelligence that leads to the realization of autonomous robots without human interference. Since agriculture is developing with the industry, this means that it is impacted by the use of IoRT to fight the problem of weeds.

ICT systems employ algorithms that can handle a greater variety of instances. However, when combined with ML, it appears to be one of the best bets for addressing new agricultural problems. Since most ICTs include upfront expenses, farmers are less likely to adopt them, especially in nations where agriculture is a major economic driver.

LPWAN technology is ideal for IoT connectivity, linking devices directly to the Internet without IoT gateways. Agriculture applications that need long-range connectivity, little power use, and close to real-time communication are well suited for LPWAN.

ACKNOWLEDGMENTS

This work was supported by the Ministry of Higher Education, Scientific Research, and Innovation, the Digital Development Agency (DDA), and the CNRST for the realization of the first Moroccan agricultural robot.

REFERENCES

[1] "Applications of IoT in Robotics—Things of IOT." https://www.thingsofiot.com/blog/applications-of-iot-in-robotics (consulté le 13 octobre 2022).

[2] "Robotics." Wikipedia, 7 octobre 2022. [En ligne]. https://en.wikipedia.org/w/index.php?title=Robotics&oldid=1114678829 (consulté le 13 octobre 2022).

[3] B. Miles, E.-B. Bourennane, S. Boucherkha, et S. Chikhi, "A study of LoRaWAN Protocol Performance for IoT Applications in Smart Agriculture." *Computer Communications*, vol. 164, pp. 148–157, déc. 2020, doi: 10.1016/j.comcom.2020.10.009.

[4] F. Zantalis, G. Koulouras, S. Karabetsos, et D. Kandris, "A Review of Machine Learning and IoT in Smart Transportation." *Future Internet*, vol. 11, no 4, Art. no 4, avr. 2019, doi: 10.3390/fi11040094.

[5] Harshith, "IoT Applications in Robotics and Automation." *IoTEDU*, 1 mai 2020. https://iot4beginners.com/iot-applications-in-robotics-and-automation/ (consulté le 13 octobre 2022).

[6] "IoT in Agriculture: Internet of Things Solutions for Smart Farming." *Digiteum*, 17 mai 2021. https://www.digiteum.com/iot-agriculture/ (consulté le 14 octobre 2022).

[7] "IoT in Agriculture: For Real-Time Farm Monitoring." https://www.cropin.com/iot-in-agriculture (consulté le 14 octobre 2022).

[8] S. F. A. Shah et al., "The Role of Machine Learning and the Internet of Things in Smart Buildings for Energy Efficiency." *Applied Sciences*, vol. 12, no 15, Art. no 15, janv. 2022, doi: 10.3390/app12157882.

[9] J. Bzai et al., "Machine Learning-Enabled Internet of Things (IoT): Data, Applications, and Industry Perspective." *Electronics*, vol. 11, no 17, Art. n° 17, janv. 2022, doi: 10.3390/electronics11172676.

[10] A. L. Rosa, "What Is LPWAN? An Introduction to the IoT Communications Protocol." *Pandora FMS Monitoring Blog*, 20 septembre 2018. https://pandorafms.com/blog/what-is-lpwan/ (consulté le 20 février 2023).

[11] F. Pasandideh, "IoT Based Smart Farming: Are the LPWAN Technologies Suitable for Remote Communication?" août 2020, doi: 10.1109/SmartIoT49966.2020.00048.

[12] F. J. Dian et R. Vahidnia, "Chapter 3: LPWAN." déc. 2020. [En ligne]. https://pressbooks.bccampus.ca/iotbook/chapter/lpwan/ (Consulté le 20 février 2023).

[13] R. Sanchez-Iborra, "LPWAN and Embedded Machine Learning as Enablers for the Next Generation of Wearable Devices." *Sensors (Basel)*, vol. 21, no 15, p. 5218, juill. 2021, doi: 10.3390/s21155218.

[14] "What Is the Internet of Things (IoT)?" https://www.oracle.com/internet-of-things/what-is-iot/ (consulté le 22 mars 2023).

[15] "What is IoT (Internet of Things)?" Microsoft Azure. https://azure.microsoft.com/en-us/resources/cloud-computing-dictionary/what-is-iot/ (consulté le 22 mars 2023).

[16] T. Team, "Architecture of IoT." *TechVidvan*, 22 octobre 2021. https://techvidvan.com/tutorials/architecture-of-iot/ (consulté le 17 octobre 2022).

[17] "Review of Agricultural IoT Technology." Elsevier Enhanced Reader. https://reader.elsevier.com/reader/sd/pii/S2589721722000010?token=9C84960C7110ECD0B68F9FD3505BC678394DBE67CE3AA8ACC625149B5A7CA1626A032BDB7325F585D4E0645172E0B07E&originRegion=eu-west-1&originCreation=20221014143453 (consulté le 14 octobre 2022).

[18] "GIS, RS and GNSS in Whatever." https://www.gim-international.com/content/article/gis-rs-and-gnss-in-whatever (consulté le 4 novembre 2022).

[19] "Types of Networks I PAN vs LAN vs MAN vs WAN vs CAN ★." *IPCisco*. https://ipcisco.com/lesson/types-of-networks/ (consulté le 4 novembre 2022).

[20] "Cloud Storage vs. Cloud Computing: What's the Difference?" https://info.cloudcarib.com/blog/cloud-storage-vs.-cloud-computing-whats-the-difference (consulté le 4 novembre 2022).

[21] "What Are Cloud Applications?" https://www.redhat.com/en/topics/cloud-native-apps/what-are-cloud-applications (consulté le 4 novembre 2022).

[22] V. Beal, "What Are Cloud Services?" *Webopedia*, 27 juin 2011. https://www.webopedia.com/definitions/cloud-services/ (consulté le 4 novembre 2022).

[23] "What Is Weeding?—Crop Protection through Weeding." *BYJUS*. https://byjus.com/biology/weed-crop-protection-weeding/ (consulté le 5 octobre 2022).

[24] "Planting Definitions I What Does Planting Mean? I Best 4 Definitions of Planting." https://www.yourdictionary.com/planting (consulté le 4 novembre 2022).

[25] T. Team, "What Is Harvesting? Crop Production & Management—Class 8 Science." *Takshila Learning*, 18 juillet 2018. https://www.takshilalearning.com/what-is-harvesting-crop-production-management-class-8-science/ (consulté le 4 novembre 2022).

[26] "An Architectural Framework Proposal for IoT Driven Agriculture | Request PDF." https://www.researchgate.net/publication/333831207_An_Architectural_Framework_Proposal_for_IoT_Driven_Agriculture (consulté le 12 novembre 2022).

[27] "7 IoT Smart Solutions in Agriculture Sector [2021]." *Ubidots Blog*, 31 décembre 2021. https://ubidots.com/blog/iot-in-agriculture/ (consulté le 14 octobre 2022).

[28] J. P. Queralta, T. N. Gia, Z. Zou, H. Tenhunen, et T. Westerlund, "Comparative Study of LPWAN Technologies on Unlicensed Bands for M2M Communication in the IoT: Beyond LoRa and LoRaWAN." *Procedia Computer Science*, vol. 155, p. 343–350, janv. 2019, doi: 10.1016/j.procs.2019.08.049.

[29] T. A. Khoa, M. M. Man, T.-Y. Nguyen, V. Nguyen, et N. H. Nam, "Smart Agriculture Using IoT Multi-Sensors: A Novel Watering Management System." *Journal of Sensor and Actuator Networks*, vol. 8, no 3, Art. no 3, sept. 2019, doi: 10.3390/jsan8030045.

[30] M. L. Liya, et D. Arjun, "A Survey of LPWAN Technology in Agricultural Field." In 2020 Fourth International Conference on I-SMAC (IoT in Social, Mobile, Analytics and Cloud) (I-SMAC), Palladam, India: IEEE, oct. 2020, pp. 313–317. doi: 10.1109/I-SMAC49090.2020.9243410.

[31] V. Meshram, K. Patil, V. Meshram, D. Hanchate, et S. D. Ramkteke, "Machine Learning in Agriculture Domain: A State-Of-Art Survey." *Artificial Intelligence in the Life Sciences*, vol. 1, p. 100010, déc. 2021, doi: 10.1016/j.ailsci.2021.100010.

[32] L. Benos, A. C. Tagarakis, G. Dolias, R. Berruto, D. Kateris, et D. Bochtis, "Machine Learning in Agriculture: A Comprehensive Updated Review." *Sensors*, vol. 21, no 11, Art. no 11, janv. 2021, doi: 10.3390/s21113758.

[33] T. Utstumo et al., "Robotic In-Row Weed Control in Vegetables." *Computers and Electronics in Agriculture*, vol. 154, pp. 36–45, nov. 2018, doi: 10.1016/j.compag.2018.08.043.

[34] "An Autonomous Robot for Weed Control: Design, Navigation and Control." *Docslib*. https://docslib.org/doc/6840144/an-autonomous-robot-for-weed-control-design-navigation-and-control (consulté le 12 novembre 2022).

Chapter 4

A study on IoT networks and machine learning using low-power wide area networks technology

Samina Kanwal
COMSATS University Islamabad

Junaid Rashid
Sejong University

Samira Kanwal
University of Engineering and Technology

Nasreen Anjum
University of Gloucestershire

Sapna Juneja
KIET Group of Institutions

4.1 INTRODUCTION

The amount of IoT-based technology is predicted to grow dramatically to more than 28 billion by 2021 [1]. There has been an increase in the use of smart and linked edge devices in our daily lives, from smart refrigerators [2] to wearable [3,4]. A smooth interaction with the application environment is essential for the Internet of Things (IoT) to continue growing and becoming more pervasive. An essential component of this intended integration is the IoT's capacity to operate autonomously for extended periods without the need for recharging [5]. In addition, the demand for IoT in more complicated applications necessitates the ability to transmit, store, and process data all of which consume a lot of power. Using the cloud [6] for data analysis purposes, where raw data is transferred for processing and analysis, has become a common technique for data processing. However, the range of cloud-based IoT is constrained since sending huge amounts of data necessitates a constant power supply or regular recharging [7]. In addition, it is

DOI: 10.1201/9781003426974-4

expensive to maintain a cloud-based infrastructure [8], which might have serious financial ramifications for consumer electronics.

The fourth industrial revolution nowadays is exposed as the time when machine learning (ML) and appliance are connected wirelessly. Low power wide area network (LPWAN) is one of the most important communication technologies employed to create communication between devices across long distances. This technique achieves low latency and low data rate operation at the tradeoff of long-range and low-power operation. It does not need a high data rate and can tolerate delay [9]. It is preferably employed LPWAN is more alluring due to its capacity to provide IoT devices with low power and widespread connection over vast geographic regions at a reasonable cost [10].

As estimated by the International Energy Agency [3] by 2020, there will be more than 20 billion network-enabled devices. The significance of LPWAN networks derives from the technological abilities required to handle many devices as Internet coverage grows [11]. Basements and other hard-to-reach interior areas benefit from improved signal propagation owing to LPWAN (+20 dB gain over standard cellular systems). It enables connections between end devices and the base station that are more than 10 km long, dependent on the deployment, and the GHz band is used by the majority of LPWAN because it enables reliable communication with a minimal power budget [12,13]. In terms of communication distance, LPWAN uses any application that needs a high coverage distance and low delay.

Numerous scenarios and applications, such as smart cities (including smart meters and parking management), personal IoT applications, tracking and monitoring of animals, and industrial asset monitoring, are anticipated to leverage LPWAN. Without the need to operate a complicated mesh network, LPWAN makes it simple to deploy smart sensors everywhere throughout the region. By directly connecting end devices to the base station, the majority of LPWAN adopt the star network architecture to save a significant amount of energy [14].

Over the past 10 years, artificial intelligence (AI) has made significant progress due to the introduction of cloud computing, GPU computing, and other hardware improvements [15]. ML is the most popular AI technique used in different fields encompassing ML, computer animation, natural language processing (NLP), and computer networks. Several research studies [16–20] examined how ML is used to handle problems with routing, traffic engineering, resource allocation, and security.

The primary technology for intelligent network management and operation is ML. As the majority of IoT systems are growing more dynamic, varied, and sophisticated, managing IoT systems is extremely challenging. Furthermore, the services offered by such IoT systems must be enhanced regarding diversity and effectiveness to draw in more consumers. Many researchers have improved the use of ML in IoT. We may therefore conclude that ML support for IoT is beneficial. The use of ML for IoT helps users get in-depth insights and create effective intelligent IoT applications. Since it may provide practical methods to extract the characteristics and data hidden in IoT. The main goal of ML algorithms is to enhance the efficiency of any activity through repetition and learning, and they are capable of analyzing big datasets in any cyber-physical system with amazing analytical skills [17,21]. Machine-to-machine communication is a significant factor of IoT that is expanding quickly at a CAGR of over 20% [22]. The market, however, is fractured and oversaturated with several communication systems and specialized services that may or may not integrate properly.

Despite the potential benefits of IoT enterprises and end users are faced with a bewildering number of access options. As a result, it becomes challenging to determine how and where to start an IoT strategic plan as a strategy for their path toward digital transformation. The lack of comparative research on LPWAN systems has been identified as the main problem for potential IoT consumers [23].

The main purpose of the study is as follows:

- The study provides a detailed discussion of the IoT LPWAN application standard and challenges. Additionally, it investigates the application, security challenges, and techniques of ML in the context of IoT.
- The focus of this study is the evaluation of Wi-Fi HaLow, LoRa, and NB-IoT technologies. These three technologies are assessed based on their potential in terms of electricity efficiency, service demand, and cost.
- This chapter includes a comprehensive analysis, and comparison of current surveys, current research issues, and potential prospects for IoT security using ML.

The other portions of the article are arranged similarly to how Section 4.2 presents related material. The use of IoT based on ML is addressed in Section 4.3. The application of LPWAN-based IoT is discussed in Section 4.4. Section 4.5 presents ML challenges, whereas Section 4.6 has LPWAN challenges. Section 4.7 identifies the conclusion and future work.

4.2 LITERATURE REVIEW

LP-WAN technologies are replacing older technologies in several use cases and are increasingly being employed as last-mile connections. However, the majority of current research compares LP-WAN technologies is whether researchers have reviewed [24–28] or quick tests using two or three different methods with a specific system-related emphasis. The studies do not properly investigate numerous performance metrics or evaluate how various technologies perform in different scenarios with the same system features. Some assessment instances only examined the technology in a single context, which limits their applicability.

Many of the studies that developed real world systems did not perform comparative analysis with other LPWAN technologies. In actuality, a single technology was employed for a specific use case in several of the tests, limiting the evaluation results to scenarios of a certain sort [29,30]. Contrasts were confined to the physical and associated MAC layers in several further papers, while some assessment studies focused on either cellular-only or non-cellular LP-WAN systems [12,22,31]. The fact that many LP-WAN technology remains in their initial stages; meanwhile, various operational issues are being addressed. As a consequence, a cross-cutting comparison will not only demonstrate a standard's applicability for particular applications but will also highlight how differently LP-WAN standards were designed in terms of network performance.

LoRaWAN is also one of the most frequently studied LP-WAN technologies in the literature and also the most popular LPWAN access method for the IoT [32]. The propagation of IoT signals is crucial [33] because it affects network coverage, dependability, and data flow, problems with coverage and path loss have been reported in several LPWAN investigations, using LoRa [34] and NB-IoT [35]. Meanwhile, following a series of measurement studies [36] suggested a LoRa model for urban channel loss further empirical research such as in [37], assessed the efficacy of the LoRa's Chirp Spread Spectrum (CSS) modulation approach using frequency shift keying (FSK). The study revealed that payload size affected network coverage as well as reliability. The experiments, however, were only conducted in an academic environment, and outcomes offered little information on possible actions in various situations. More challenging terrains with higher levels of interference, including congested urban areas and other metropolitan settings, may function differently as a result of variations in the interference sources and patterns. In addition, providing fresh information will make it simpler to analyze the results in light of each unique field condition. A study [38] examined the LTE-M and NB-IoT capacity and coverage for cellular LP-WANs in a rural Danish location. According to the study, cellular technologies such as LTE-M can sustain the greatest coupling loss of roughly 156 dB, while deep interior applications may suffer an additional 30 dB of penetration loss. A feature that compares cellular technology to non-cellular technology would be fascinating. According to the study, cellular technologies such as LTE-M can sustain a maximum coupling loss of roughly 156 dB, while deep interior applications may suffer an additional 30 dB of penetration loss. A feature that compares cellular technology to non-cellular technology would be fascinating.

IoT aims to understand human behavior and mental processes, foresee desired and unwanted consequences, and improve situational management abilities. To accomplish these, IoT must interpret the data created by millions of devices. ML plays a critical and vital part in the IoT paradigm. IoT is ubiquitous by nature, therefore being accessible everywhere is one of its main objectives [39]. IoT devices will become more useful as a result of ML, and only truly widespread IoT will be possible [40]. To allow IoT to significantly contribute to the accomplishment of its goals, embedded intelligence will be at the center. To improve automation, efficiency, productivity, and connection, EI fuses product and intelligence [41,42]. Learning is the only way to become intelligent, whether in the real world or a virtual one.

The proclivity of ML to seek design might be the foundation for intellect. We may get a deeper understanding of the world around us when these patterns are generalized into more insightful insights and trends. The fundamental goal of ML in IoT is to provide entire automation by increasing learning that supports intelligence through smarter things [43,44]. After learning from data, ML allows IoT-enabled devices to mimic human-like judgments and improve their understanding of their environment. Information visualization has a major influence on the human visual system, increasing how information and insights are understood by systems [43]. Users of information visualization benefit from some factors, including (1) improved understanding without requiring extensive data analysis and (2) the use of cognitive abilities to help people understand data. Several currently utilized systems that are expensive to install and maintain will be replaced by IoT with low-cost sensor-based ML systems. For instance, extreme weather conditions caused almost 20,000 individuals to pass away in poor nations. Most often, radar-based weather monitoring systems are used to monitor the weather (WMSs). But in many places throughout the world, radar WMSs are expensive and inaccessible. In economically underdeveloped nations like Guinea and Haiti, an ML-enabled IoT system with a low-cost sensor network reliably predicts the weather by analyzing illumination and cloud patterns [11].

IoT will influence how we perceive technology as well as how it advances and improves the state of the planet [45,46]. Every day, the IoT makes many aspects of our lives more convenient and linked. Due to ML, IoT has become intelligent and ubiquitous. The term "heterogeneous" is well suited to describe the IoT, which is made up of a wide range of hardware, Internet protocol, data types, software, and users. The diverse structure of IoT provides ML with a variety of challenges, including IoT will create large amounts of data [47–49], but are all of the data legitimate, do they include biases, and is it useful to manage them all? A variety of critical factors impact the dependability, precision, and efficacy of MLAs for the IoT domain are:

- Not all IoT applications provide a large amount of data from which MLAs can learn fast instead, numerous little amounts of data are generated for this new class of algorithms required to learn from sparse data [50].

- The accuracy and dependability of sensing equipment vary [51–55]. Among the activities that must be completed on the data before ML commences are outlier detection and data imputation.
- Peter Norvig, Google's head scientist, is credited with saying, "We don't have better algorithms than anyone else; we just have more data," during the Zeitgeist 2011 conference. Few researchers, nevertheless, back the other view. What is superior? Sophisticated.
- MLA [44], with more data [56–58], or fewer data but of higher caliber [59,60]. Since there is no consensus on this issue, it is one of the major points of contention among ML researchers.

Game-changing technologies like IoT provide organizations with a wealth of options, but they also carry significant risks, and ML-enabled IoT may wind up eliminating numerous jobs [61,62].

ML provides IoT-empowered systems with a brain to extract insights from the data generated by millions of IoT devices. IoT employs a range of MLAs that learn from different types of data, which sets it apart from ML. There will be several types of MLAs, and some of them won't require complex mathematical proofs to work [44]. Early in 2001 [63], researchers published an intriguing study showing that learning is improved by more training data rather than by generating more MLAs. The Internet-connected IoT gadgets, which number in the billions, will produce enormous volumes of data [64]. IoT is not just about large data; it also involves little data [50], hence the IoT domainis only partially correct. Small data sets only have a few properties. Small data is a byproduct of the aggregation of large data and may be used to explain the present status, start events, and create small data. A wide range of IoT-enabled applications that use ML is available to governments, businesses, and consumers. MLAs and the IoT were investigated as potential applications [65]. The focus of the research was "machine learning as a service" (MLaaS), which is the combination of ML and IoT infrastructure [66]. This chapter examines historical and contemporary developments in IoT application layer protocols, as well as pertinent real-time applications and the application layer protocols they have modified for enhanced performance [67]. The artificial neural network and attribute-based tree structure used in this study give a solution to the trust difficulties addressed in Lagrange's-PI (LI). To create a durable network system, the suggested structure increases not only the trust factor but also the accuracy measurements of LI [68]. A technique called Fog-Cluster-Based Load-Balancing with a refresh period was suggested to maximize the usage of all the resources in the fog sub-system. In comparison to current methods for the proposed framework, the simulation results demonstrate that "Fog-Cluster-Based Load Balancing" reduces energy consumption, the number of Virtual Machines (VMs) migrations, and the number of shutdown hosts [69]. This study examines coexistence problems in wireless body area networks (WBANs), which are homogeneous network environments. A network specifically created for the human body is IEEE 802.15.6. Three distinct types of interference-avoiding strategies are provided by IEEE 802.15.6. These methods do not offer a sufficient means of preventing cohabitation [70].

4.3 APPLICATION PERSPECTIVE OF ML

The definition of IoT in this study [71] is no longer accurate. It is now seen to be a finding with the potential to alter the course of human history in a similar way to how electricity did. To provide readers with a clear knowledge of IoT advances in industries, the studies systematically present a comprehensive assessment of IoT industry challenges, R&D trends, and current application areas. Data are as crucial to the IoT as electrons are to electricity [72,73]. This section examines IoT ML advances and categorizes IoT applications.

Almost 50% of people on Earth live in cities, according to a United Nations survey, owing to improved employment, education, medical, and housing conditions, putting tremendous strain on cities, and governments to supply adequate resources. As a result, the "smart city" idea has lately received substantial attention from governments throughout the world, particularly in industrialized [74,75] and emerging economies [44]. Smart cities are increasingly an important component of the strategy for urban growth. A smart city does not have a specific definition. However, it may be characterized as a result of rapid progress and improved information technology, to improve residents' socioeconomic situations and general quality of life. Figure 4.1 illustrates the applications of ML based on IoT.

4.3.1 Smart grid

By eliminating power waste, reducing defects, and improving the power supply to satisfy the increased demand for electricity, smart grids increase energy availability and efficiency provide to cities, towns, and enterprises with a continuous power supply [44,76]. Even though power-grid failures are infrequent, they can cause blackouts, social unrest, and a loss of millions of dollars. Power is delivered through smart grids in a more dispersed, flexible way. IoT makes it possible to rely on smart grids for improved power management. Randal Bryant et al. predict that ML will make a significant and above-average contribution to the success of smart grids. It explains how MLAs are accelerating the development of the "data to knowledge to action" paradigm in the energy-space domain [77].

In refs. [78–80], the authors looked at some of the MLAs that are often used to assess how well they facilitate operational choices in smart grids.

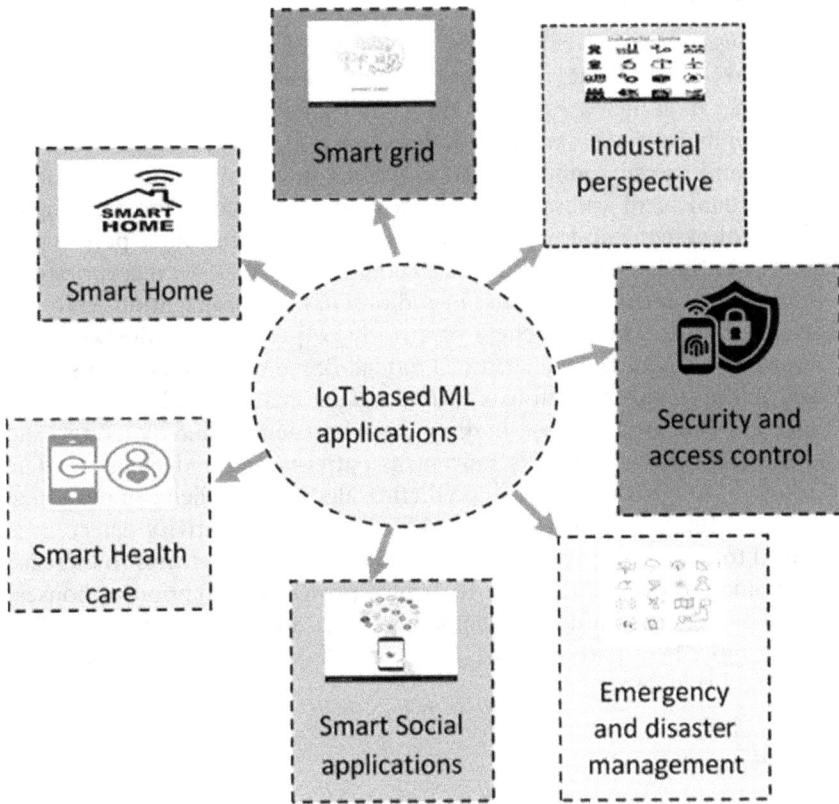

Figure 4.1 IoT-based ML application.

BP [81,82], radial basis function [83,84], multilayer perceptron [85], and optimized and hybrid ANNs [86,87] are some of the ANN types that are utilized for demand forecasting. With less overfitting, EDF by SVM [88–90] can manage noise well. ANNs and SVMs perform exceptionally well with traditional electricity demand forecasting (EDF). A revolutionary deep-learning model exhibits a notable improvement in prediction accuracy [91]. For the IoT, ANNs, and SVMs are computed costly. EDF relies on dynamic elements of conventional grid data; IoT networks incorporate weather, social events, individual interests, authority performance, and maintenance. The future of human urbanization depends on smart grids, but managing them is crucial. With ML-enabled IoT, this is achievable. The security of electrical grids connected by IoT technology, however, faces several difficulties as a result [92].

4.3.2 Smart homes

Smart houses with IoT capabilities are a technological idea that makes it possible to fully automate the functioning of household equipment and gadgets online. As it increases user comfort and safety, context awareness is a crucial component of smart homes. But there is less direct contact between the user and the surroundings. To make the home environment respond like a rational agent and save running costs, the MavHome project couples multi-agent systems and probabilistic MLA [93]. As customer preferences change over time, a more sophisticated context-aware model incorporates a reinforcement learning algorithm in addition to back-propagation ANN for service selection. The key benefit of a study [94] is that the context-aware system doesn't need a predetermined model. Based on the customer's comments on the service, modeling is carried out automatically. SHs are capable of logical automated decision-making. This is accomplished by monitoring and forecasting the resident's movement patterns and gadget usage. The Markov chain-based active LeZi prediction algorithm, which can recognize patterns in future events, is proposed in [95]. Human activity detection is a crucial topic that has lately received a lot of interest in efforts to increase the automation of SHs (HAR). To reduce power consumption in houses, the author suggested a deep-learning algorithm for motion categorization utilizing movement patterns [96]. However, the computational cost of the DL method is high.

4.3.3 Smart healthcare

The IoT is bringing out new, sophisticated sensors that are connected to the Internet and provide vital data in real-time, transforming the healthcare sector. In-depth explanations of IoT in healthcare, platforms, applications, and market trends for smart healthcare solutions were provided [97]. Smart healthcare apps aim to: (1) enhance and simplify access to care; (2) raise the standard of treatment; and (3) lower the cost of care. Perceiving patterns and important insights from healthcare data is essential for attaining the aforementioned goals [98,99]. These studies [100] offer suggestions for managing weight, monitoring cardiovascular disease, and monitoring physiological data. In [101], a wearable armband multisensory system called Body Media FIT uses ML to continuously measure physiological data and regulate body weight. The technology, which employs regression analysis to categorize activities, has been commercially accessible since 2001. Mobile devices are used to track cardiac conditions. By contextualizing vital health sign patterns with clinical data sets, M4CVD analyzes them locally [100]. A comprehensive intelligent healthcare system with IoT capabilities that caters to a wider range of patients is offered by IBM Watson. It blends the strength of MLAs with the capacity of healthcare data to provide fresh insights [102]. Cognitive care offers up-to-date ways for medical professionals to communicate with their patients, enhancing diagnosis accuracy and lowering mistake rates.

4.3.4 Smart social applications

IoT has greater potential to impact social elements of human existence than we can conceive, aside from the technology aspect. ML-enabled IoT may be used to analyze social application data for patterns and determine the public's sentiment toward a certain topic. Utilizing ML, views may be created and public perceptions can be studied using connected devices such as cell phones and tablets. This research presents a complete evaluation of all potential methodologies, namely ML, that may be challengers, as well as the most current developments in the field [103]. By examining popular feelings, opinions can be anticipated. Based on unique data formats and MLAs, the authors of [104] suggested a sentiment analysis approach that can interpret the emotional orientation of Arabic Twitter postings. The suggested method made extensive use of lexical, surface-form, and syntactic properties. They also made use of lexical characteristics extrapolated from two Arabic lexicons of sentiment words. To create a supervised sentiment analysis system, the authors used several conventional categorization techniques. The MLAs for feature extraction and the subsequent categorization of SVM and NB are used to identify only pertinent content in the tweets. Further, they illustrate how social media analysis may currently be utilized to identify dangers and undesirable occurrences [105].

4.3.5 Emergency and disaster management

We can intensely improve our ability to expedite relief operations during any emergency or disaster by using IoT technology. The IoT paradigm was thoroughly investigated by this study [106] with an emphasis on disaster management, from its application domain through ML-based advanced analytics. An early forecast of a forest fire is one of the essential applications that can benefit from IoT technology. For instance, reaction time must be considerably reduced in the case of forest fires spreading so fast. However, because of outliers, an IoT-based system may forecast the incorrect event. To address this issue, we suggested an IoT architecture that utilizes classification and regression [107]. The researchers classified current unsupervised machine learning techniques [108]. In terms of human fatalities and property damage, we have seen the devastation brought on by tsunamis during the past 10 years. This study [109] developed a tsunami early warning system, which solved the issue. An RF classifier uses seismic data to forecast the presence of a tsunami.

4.3.6 Control accessibility and smart security

In the IoT context, it is important to protect the confidentiality and safety of millions of people and even billions of users since sensitive information is captured and supplied to their applications [110]. Several access-control methods based on classification, clustering, and reinforcement learning were used in research [111]. A more recent state of art thoroughly examined many assaults, cutting-edge security concerns, and state-of-the-art solutions. There were several significant security holes discussed. As a security measure for the IoT paradigm, the authors suggested expanding machine learning approaches, which are limited to building intelligence [112]. In a study [113], numerous ML techniques, including decision trees, random forests, support vector machines, logistic regression, and artificial neural networks are used to predict different sorts of assaults and abnormalities in the IoT context (ANN). A thorough evaluation of biometric-based access-control methods in [114] included, artificial neural networks (ANNs), linear discriminant analysis, learning vector quantization, and Fisher discriminant analysis. Their merits and downsides were also covered. It's interesting to note that the method in [97] only makes use of little computational power.

4.3.7 Industrial perspective

The information technology (ITI) sector's development and use of IoT are still in their early phases. ITI is gradually but increasingly seeing the potential value of IoT. A lot of R&D is being conducted, motivated by hopes, market trends, and data. The development of new IoT-enabled software platforms and hardware is being spearheaded by ITI heavyweights such as Google, IBM, Oracle, and SAP. IoT infrastructure adoption will be considerably accelerated and improved by expanding its use, which will transform industry operations [115]. Finding important insights, or more simply, making sense of data created by the IoT is one of the largest challenges facing the IoT. ML can address these problems. ITI begins by adding MLAs as they begin to gather more data for their IoT-based systems in light of these crucial facts. Microsoft built the cloud computing system known as Azure [116]. IoT is generating enormous volumes of data from millions of endpoints each day, and the latest changes in Azure are a consequence of his remarks. Microsoft Azure's main program manager, Scott Hanselman, gave an example of combining several components and making ML for IoT possible [44]. Another intriguing advancement was made by IBM with the Watson software platform [44,117], which was created for Jeopardy Answers. A technological platform called IBM Watson employs ML and NLP to extract insights from enormous volumes of unstructured data. For instance, a car owner would wish to know when a certain auto part should be scheduled for predictive maintenance. Watson accomplishes this by using sensor data that has been accumulated over time to analyze the performance of the machinery and the breakdown time.

This innovation demonstrates IBM's readiness to use edge analytics to reduce redundant data exchange to the cloud. The efforts of Cisco in fog computing will be very helpful in disseminating intelligence at the edge. In this relationship, Watson's job is to supply a little bit of code that alerts the program to the fascinating facts relevant to a certain need. TensorFlow (TF), an open-source ML framework, was released by the search engine behemoth Google. TF is currently used by some Google products. For instance, TF is used by voice recognition, email, Google pictures, and search. The fact that TF is extremely scalable and can run on a variety of platforms, servers, personal computers, cell phones, and other mobile devices is a significant benefit. Users can run customized MLAs. By employing TF, deep learning's potential may be fully realized. Splunk, a US-based global firm, launched IoT-enabled ML software called Splunk (product), which is great at drawing general conclusions from operational data. It effectively manages huge data and enhances maintenance and problem-finding using data provided by the IoT [118]. There are almost 300 new MLAs in total [49]. Splunk emphasizes the advantages of its ML system for non-technical users. Splunk interfaces with well-known IoT platforms and services, which may contribute to Splunk's increased popularity. There are also additional businesses on the market that provide IoT application-specific ML solutions. IoT is a huge application space for ML. Unquestionably, IoT's possibilities and problems [53,119] are considered ITI's interest in creating ML-enabled IoT.

4.4 APPLICATION PERSPECTIVE OF LPWAN

The idea of the IoT has now been expanded. It will have a significant impact on how humanity approaches its current problems in the coming years. The key areas of distinction among LPWAN applications are healthcare, smart cities, e-learning, smart grid work, and environmental safety.

4.4.1 Intelligent power network

In recent years, efforts have been undertaken to switch to renewable energy sources. Due to the unpredictability and geographical spread of all of these sources, remote monitoring and information transfer will be required.

4.4.2 E-learning

IoT must first make education more accessible. The linked WSN networks that would enable long-distance information transfer would make it feasible for anybody to access distant learning.

4.4.3 Smart city

One of the benefits of IoT is the more effective utilization of resources. Several articles have examined the effects of WSN technology on intelligent cities.

4.4.4 Healthcare

WSN technologies allow for the monitoring and protection of patients while they are at home, enhancing the standard of care.

4.4.5 Environment safety

The preservation of the environment is a pressing issue that cannot be ignored. Remote monitoring of pollution levels and the effects of human activity on the environment is made possible by WSN sensor networks.

4.4.6 Standards and techniques

Systems that can tolerate delays, don't need a lot of data transfer speed, need little power, and need a wide area of coverage are good candidates for LPWAN technology. Figure 4.2 depicts an IoT-based LPWAN standard and techniques.

4.4.6.1 Low power

The duty cycle and the network architecture are only two of the many variables that affect battery life.

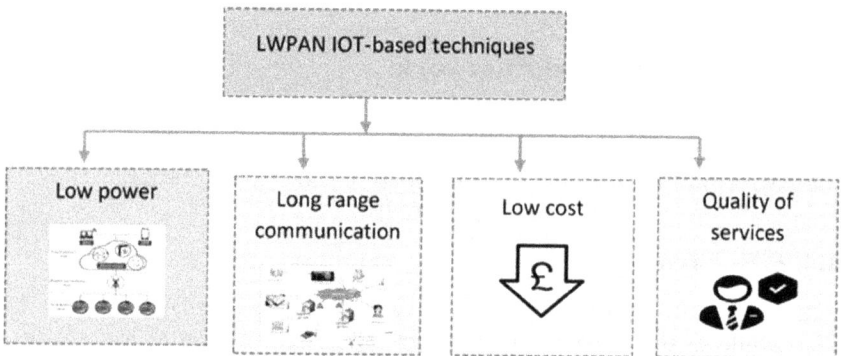

Figure 4.2 IoT-based LPWAN techniques.

Topology: IoT devices should last for a very long period. Therefore, a battery lifespan of at least 10 years is preferred to save maintenance costs. Although mesh topology is employed to broaden short-range wireless networks, its high implementation costs make it difficult to link several end-user devices over significant distances. The network lifespan is just a few months or years when there is traffic over several hops because some nodes become overcrowded, and their batteries run out. Because of this, the majority of LPWAN adopt start topologies, which cut off the energy used by packet multichip networks [9] and link devices directly to the base station, resulting in significant energy savings.

Duty cycle: Devices are turned off when no data is being delivered to save power consumption. LPWAN duty cycle is customized for each application, power source, and volume of traffic. Only when data transmission is required is the transceiver turned on. When a program requests data transfer in uplink mode, end devices may only awaken if the data are prepared to be transmitted. This knowledge is reached by the end devices by choosing a listening schedule [120].

4.4.6.2 Long-range communication

According to the deployment context (country or urban), LPWAN provides superior signal transmission to distant interior locations like basements enabling end-to-end device connectivity across tens to hundreds of kilometers worth of space [9]. The phrase "wide area" describes a range that may be directly covered without the requirement for a mesh network [14]. Sub-GHz and unique modulation methods can be employed to reach extended distances.

The majority of LPWAN utilize sub-GHz due to its dependability, strong communication, and low power budget. It has less frequency attenuation than 2.4 GHz Sub-GHz and less multipath fading and congestion caused by barriers and thick surfaces like concrete walls. Sub-GHz achieves lengthy communication and low power consumption because of its durability and excellent dependability. Specialized modulation techniques: LPWAN with a link budget of around 150 dB, designed to cover a range of a few to tens of kilometers. Different LPWANs use the two strategies known as narrowband and spread techniques.

With a limited bandwidth (less than 25 KHz) for the signal encoding, narrowband, such as NB IoT, offer a high link budget. A narrowband signal is spread using the spread spectrum technique with the same power density over a broader frequency band [120]. Wi-Fi HaLow's projected coverage distance is 1 kilometer, which is less than that of NB-IoT and LoRa (10 km). Due to its usage of the LTE network, which has a wider radio coverage range than LoRa and Wi-Fi HaLow, NB-IoT often provides superior coverage distance.

4.4.6.3 Low cost

One of the main factors determining LPWAN's success is cost, so the technology employs several strategies to cut costs, including switching to a star topology instead of a mesh topology, reducing hardware complexity using a minimal infrastructure, and using license-free or own bands. A few LPWANs use the cellular band to cut costs on licensing. Due to the higher cost of the spectrum, LoRa and Wi-Fi HaLow frequently have lower total costs than NB-IoT. Since LoRa is unlicensed and Wi-Fi HaLow is not required to have a license, their combined cost is cheaper. The cost of the spectrum for LoRa and Wi-Fi HaLow is nothing, while it may be predicted to be 500 million USD/MHz for NB-IoT. While LoRA and Wi-Fi HaLow only cost around $1,000 per station to deploy, NB-IoT is anticipated to cost $15,000 per station [121].

4.4.6.4 Quality of service

LPWAN is utilized in several applications, each of which has different needs. While certain applications, like smart meters, can accept transmission delays, others cannot. As a result, the quality of service (QoS) relies on how an application is used. Given that both NB-IoT and Wi-Fi HaLow are based on the licensed LTE network, which is a cellular network optimized for the best QoS, and that both are wireless technologies, LoRa's use of unlicensed spectrum, CSS modulation, and an asynchronous MAC protocol do not give it a competitive advantage in terms of QoS. HaLow has the potential to be integrated with the Wi-Fi network, delivering an advantage through its strong signal penetration and MIMO capability such as mounting the sensor out of sight.

4.4.7 Technical comparison

A thorough explanation of the specifications of the three LPWAN methods under consideration is provided in this section. Figure 4.3 provides a technical comparative illustration.

4.4.7.1 Wi-Fi HaLow

Because Wi-Fi technology uses too much power, the Wi-Fi Alliance developed the IEEE 802.11 ah Wi-Fi HaLow standard, which can work in bandwidths without a license. For IoT devices, using this technique, a sizable coverage area may be achieved with little power usage. To increase the number of supported devices and the area coverage distance, several MAC

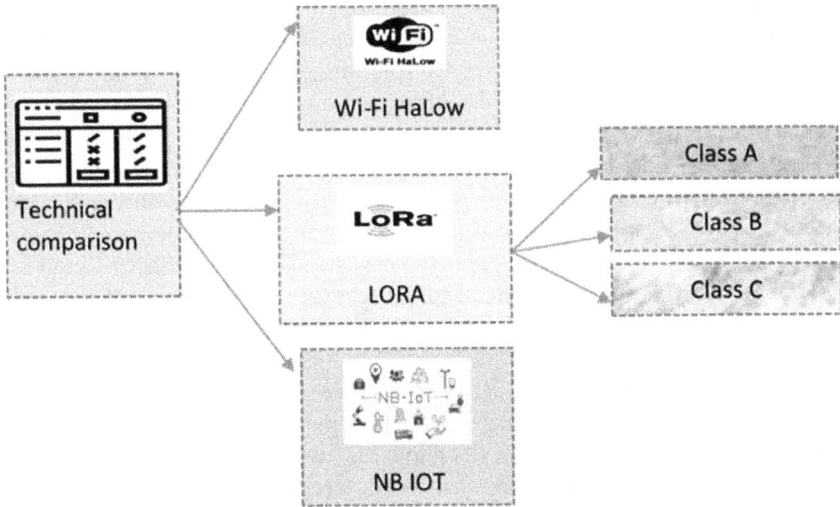

Figure 4.3 Comparative illustration.

functions have been introduced. An access point (AP) can support 8191 devices at once using Wi-Fi HaLow [122]. HaLow uses the same Sub-GHz frequency spectrum as the majority of LPWAN Wi-fi and can broadcast signals up to 1 kilometer away in open spaces. Its data rate is higher than LoRa and NB-IoT. It reaches at least 150 Kbps [123].

Wi-Fi HaLow distinguishes between TIM mode and non-TIM mode as its two operational modes. The terms "TIM station" and "non-TIM station" refer to stations that operate in one of these two modes; TIM stations periodically access the medium, are often utilized for large bandwidth needs, and get access to both the downlink and the uplink. Periodically, TIM stations awaken to receive data transmitted by the AP.

A Restricted Access Window (RAW) method was included by the standard to lessen collisions and interference. This system enables the booking of a specified time frame for a certain station. RAW is therefore used to limit access to a certain set of stations.

4.4.7.2 LoRA

CSS modulation is used at the PHY layer of Lora, a proprietary spread spectrum technology. Long-range and resistance against interference are two characteristics of CSS modulation. Scalable in terms of frequency and bandwidth is LoRa modulation. Because of its wide bandwidth and asynchronous mode, the LoRa signal is especially resistant to both in- and out-of-band interference. Lora is perfect for metropolitan and regional areas since it provides immunity to multipath and fading. Depending on the spreading factor and channel capacity, LoRa can carry data at a maximum rate of 50 kbps. Furthermore, a LoRa base station may simultaneously receive messages sent using various spreading factors [124]. Each message's payload can only be a maximum of 243 bytes. The LoRa-Alliance standardized the LoRaWAN communication protocol in 2015. Although it is a cloud-based MAC layer protocol, it also functions as a routing protocol to control communication between the LPWAN gate and end node devices. All base stations within the range receive every message that an end device employing LoRa transmits. Three kinds of devices are offered by LoRaWAN to accommodate the needs of several applications.

Class A: (bi-directional end devices): Two brief downlinks receive windows following each uplink transmission in this class. Battery-operated Class A gadgets are created largely for energy efficiency. Except when there is data to broadcast, the gadget is always in sleep mode.

Class B: (bi-directional endpoints with predetermined receive times): Battery-powered devices, which are classified as Class B are primarily intended to serve as actuators. For equipment in this class, energy efficiency is still a consideration, although not to the same extent as for class A devices. To let the server know that the end device is listening, it prepares a synchronized beacon.

Class C: (optimum receiving slot for bi-directional end devices): Class B devices and these actuators have certain similarities, except that they are powered by a constant source. Class C devices can afford to listen constantly but consume more power than class A and B devices while providing the lowest latency for end-device connection.

4.4.7.3 NB IoT

In June 2016, the 3GPP published the NB-IoT LPWAN technology, which is used to perceive and acquire data for IoT low data rate applications. Because NB-IoT operates on a licensed spectrum and has the benefit of cellular solutions over other LPWAN technologies, it is not subject to duty cycle restrictions. It offers a significant deal of promise to fulfill the IoT's communication needs for machines [12]. Three deployment options are available: Guard band, in the band, and stand-alone.

In the case of stand-alone mode, the GSM frequency band with a 200 KHz bandwidth and a guard interval of 10 KHz on either side may be used.

The unused resource block is utilized in the guard band of the LTE carrier in guard band mode.

A mode or situation that coexists with the band utilized by the LTE carrier is referred to as being in-band. The present LTE infrastructure alone won't support NB-IoT. Further software changes are required. The comparison of the three standards is shown in Table 4.1.

Table 4.1 Comparative features of the standard [125]

Variable	Lora	NB-IoT	Wi-Fi HaLow
Frequency range	868	868	900
Cellular device power	10,000	150,000	8,191
Speed of data	0.3–50 kbps	0.3–4,000 kbps usual 200 kbps	300 Mbps to 150 kbps
A lifetime of battery price	Up to 10 years	Less time, but up to 10 years, than what LoRa and Wi-Fi HaLow now provide	about 10 years
Rate	$100–1,000 per BS	Need to upgrade the base station. The BS is expected to cost $15,000 in total	$100–1,000 per BS
Text limitation	1%–10%	-	2%
Text limitation	Low	High	Medium

4.5 CHALLENGES FOR IoT-BASED LWPAN

In the IoT sector, LPWAN is expanding quickly. Both licensed and unlicensed LPWAN technologies are being developed. Here are some difficulties that LPWAN technology confronts.

4.5.1 Scalability

One of the most important characteristics of Loera, Wi-Fi HaLow, and NB-IoT is the capacity to support a large number of devices. The performance of end devices declines exponentially as their quantity increases. Scalability across large areas is thus quite difficult. Therefore, several approaches are being examined to address this scaling issue, including the usage of the ALHOA protocol. More research points to the need for end devices to adapt to LoRa communication settings, maybe with assistance from strong base stations exploiting base station diversity [120].

4.5.2 Frequency band and interference management mitigation

LoRa and Wi-Fi-HaLow, which operate in unlicensed bands, have significant challenges due to their radio spectrum limitations. The total output broadcast in an unconstrained band is 10–20 times lower than 500 mW in certain places (Russia), although this is still the case in most of Europe. The maximum power radiated is roughly 500 mW. This results in a smaller coverage area and a weaker signal, which makes implementation challenging due to the frequency band's restrictions. Due to the growing number of IoT devices, equipment operating in the common ISM band will experience unusually high levels of self-interference and interference from other technologies. Adaptive transmission scheduling spanning location, time, and frequency is recommended by certain research to reduce interference and obtain the highest dependability. Another approach to dealing with this problem is to put up regulations that would allow effective sharing and collaboration among various wireless technologies operating in unlicensed bands.

4.5.3 High transmission data rates and modulation methods

These standards do not place a high priority on transmission at a high data rate since it is frequently given up to achieve other requirements like the optimum range and power use. However, it is crucial for some applications that need high-speed, high-quality video and blood pressure monitoring. The fastest data rate is offered by Wi-Fi HaLow, which can reach 347 Mbps (with a minimum of 150 kbps). Compared to LoRa, NB-IoT can transmit data at a maximum rate of up to 4 Mbps (50 kbps). Since LoRa is primarily made for sensor networks that do not need to send a lot of data, its maximum transmission data limit range between 0.5 and 50 kbps. Using various modulation schemes for a single IoT device so that it may switch between them based on the demands of the application is one research solution method to achieving long-range, high energy efficiency, and high data rate performance at the same time [120].

4.5.4 Interconnection

The biggest difficulty this notion faces is ensuring interoperability between the products made by various vendors. Because of the many communication routes and associated applications, using a single communication protocol is not a practical option. Hence, ensuring the global interoperability of the various data formats is the major objective. Analyzing the current standards employed in the IoT idea is crucial as well.

4.5.5 Effective resource management

As the majority of WSN modules are battery-powered, there is a continuing need to lower energy usage. While the price of these gadgets is very modest, changing the battery cannot be regarded as cost-effective because it would be more expensive to replace the complete module. Hence, the battery life cycle should last several decades. This requires the creation of sophisticated communication plans with an energy efficiency focus. The development of energy-collecting techniques is a different option that may be considered.

4.5.6 Security

It is unquestionably important to consider the availability of strong security standards. Thus, it is imperative to create new authentication and encryption methods that can make efficient use of the hardware resources unique to WSN nodes.

4.5.7 Complexity

Communication capabilities integration is a very difficult process. Researchers are now working on novel approaches to boost these modules' productivity and make programming easier.

4.6 CHALLENGES FOR IoT-BASED ML

The IoT paradigm is an excellent option for bridging the distance between the real and virtual worlds by creating a hyper-connected environment. Nonetheless, it must overcome and handle tremendous technical and non-technological problems. Based on multiple IoT literature evaluations [126,127], we recognized four types of challenges: (1) technology, (2) individual, (3) business, and (4) societal.

4.6.1 Technical difficulties

The adoption of IoT systems on a broad scale is still in the future. And still, they are only used rarely in various regions of the world. The IoT is made up of four core technologies: sensors, networking, actuators, and security (IoT). There are still connection problems throughout the world. Internet accessibility and mobile connection are challenges, especially in developing nations. The client-server architecture that now dominates the IoT landscape is not a viable solution for the future due to increased latency, a higher small instance of failure, energy use, and security vulnerabilities. Edge and fog computing platforms were created to address this issue, albeit the majority of them are still in the early stages and encounter issues with network capacity, delay, accessibility, management, and control [128]. The security of these "things" is a major issue that poses a danger to the widespread adoption of IoT systems [127]. Decentralized edge devices, which are more vulnerable to hackers, are what the IoT is all about. There are additional security concerns due to the lack of universal standards for authentication and authorization for IoT devices. One of the most important concerns that IoT must answer to become a successful and accepted paradigm is security.

4.6.2 Individual difficulties

IoT will be used by various people for a range of needs. The objective is to enhance daily living through advanced technology and a sophisticated environment. Due to the "anytime and everywhere" nature of IoT, privacy is the main problem since it differs from person to person, creating increasingly complex scenarios that devices and services must handle. A study critical discussion of developing IoT trends and privacy preservation techniques. The public's acceptance of IoT is another barrier to its success. IoT will alter the way we view daily life [129]. How well these alterations align with our cognitive demands is the key consideration [130]. According to this report, public support is important for smart cities, which are fundamentally reliant on IoT architecture to maintain sustainability. To raise public acceptance of IoT, we must inform and instruct people about the benefits that these cutting-edge programs and services provide as well as how they may transform our daily lives and bring about more wealth [131].

4.6.3 Business challenges

Because of IoT, there is enormous economic potential in technologies, operations, and production. Nonetheless, it falls short of the excitement that was produced. Companies are struggling with obstacles such as the lack of network interoperability, standardized industrial platforms, a lack of global systems, data collecting, and safety concerns, among others [132]. One significant challenge is the lack of IoT professionals with the necessary skills.

4.6.4 Society challenges

We require a profitable and sustainable living environment as a civilization. IoT can considerably assist by offering assistance for actionable decision-making [133]. In a while, the critical question is: As a society, are we prepared to embrace IoT today? And are we willing to adapt to the cognitive shifts it will bring about in our day-to-day activities? The answers to these crucial queries will determine how widely IoT services and applications are accepted [134]. It will boost demand, pushing businesses to invest seriously in the widespread implementation of IoT, its goods, and applications. Another big problem that our communities deal with is the digital divide in the modern era as a result of the uneven rate of global economic progress. We need to overcome the challenges raised above if we want to fully exploit IoT.

4.7 CONCLUSION AND FUTURE WORK

IoT has now become much more developed, and its applications are being used in a wider range of settings. People, businesses, and governments all show great interest in the potential applications of IoT. In this study, instead of conducting a traditional literature analysis, we aimed to emphasize the value of ML for the development of various IoT applications. We compared three LPWAN technologies (NB-IoT, LoRa, and Wi-Fi HaLow) in terms of their key features. Through this work, the reader can understand the implications of ML for IoT, how ML is employed with IoT, and the potential of ML in IoT. Each of the three LPWAN technologies has its advantages and disadvantages, depending on the types of applications considered. The study has shown that there are still issues that need to be overcome to achieve the design objectives of LPWAN, regardless of technological advancements. From the perspective of ML, this development will pose challenges due to device data and processing capabilities. We conclude that, for the time being, IoT developments will rely on widely accessible and well-proven ML techniques. However, the development of a fully autonomous IoT environment with integrated intelligence is something we may anticipate in the future.

REFERENCES

[1] L. Mora, M. Deakin, Y. A. Aina, and F. P. Appio, "Smart city development: ICT innovation for urban sustainability," *Encyclopedia of the UN Sustainable Development Goals: Sustainable Cities and Communities*, 2019, pp. 1–17.

[2] A.-D. Floarea and V. Sgârciu, "Smart refrigerator: A next generation refrigerator connected to the IoT," In *2016 8th International Conference on Electronics, Computers and Artificial Intelligence (ECAI)*, 2016, pp. 1–6: IEEE.

[3] M. Magno, M. Pritz, P. Mayer, and L. Benini, "DeepEmote: Towards multi-layer neural networks in a low power wearable multi-sensors bracelet," In *2017 7th IEEE International Workshop on Advances in Sensors and Interfaces (IWASI)*, 2017, pp. 32–37: IEEE.

[4] M. Haghi, K. Thurow, and R. Stoll, "Wearable devices in medical internet of things: Scientific research and commercially available devices," *Healthcare Informatics Research*, vol. 23, no. 1, pp. 4–15, 2017.

[5] D. Thomas, R. McPherson, G. Paul, and J. Irvine, "Optimizing power consumption of Wi-Fi for IoT devices: An MSP430 processor and an ESP-03 chip provide a power-efficient solution," *IEEE Consumer Electronics Magazine*, vol. 5, no. 4, pp. 92–100, 2016.

[6] P. P. Ray, "A survey of IoT cloud platforms," *Future Computing and Informatics Journal*, vol. 1, no. 1–2, pp. 35–46, 2016.

[7] Z. Dawy, W. Saad, A. Ghosh, J. G. Andrews, and E. Yaacoub, "Toward massive machine type cellular communications," *IEEE Wireless Communications*, vol. 24, no. 1, pp. 120–128, 2016.

[8] A. Greenberg, J. Hamilton, D. A. Maltz, and P. Patel, "The cost of a cloud: Research problems in data center networks," *ACM SIGCOMM Computer Communication Review*, vol. 39, no. 1, pp. 68–73, 2009.

[9] U. Raza, P. Kulkarni, and M. Sooriyabandara, "Low power wide area networks: An overview," arXiv 2016, arXiv preprint arXiv:1606.07360.

[10] Z. Qin, F. Y. Li, G. Y. Li, J. A. McCann, and Q. Ni, "Low-power wide-area networks for sustainable IoT," *IEEE Wireless Communications*, vol. 26, no. 3, pp. 140–145, 2019.

[11] N. S. Knyazev, V. A. Chechetkin, and D. A. Letavin, "Comparative analysis of standards for low-power wide-area network," In *2017 Systems of Signal Synchronization, Generating and Processing in Telecommunications (SINKHROINFO)*, 2017, pp. 1–4: IEEE.

[12] J. Chen, K. Hu, Q. Wang, Y. Sun, Z. Shi, and S. He, "Narrowband internet of things: Implementations and applications," *IEEE Internet of Things Journal*, vol. 4, no. 6, pp. 2309–2314, 2017.

[13] N. Tsavalos and A. Abu Hashem, "Low power wide area network (LPWAN) technologies for industrial IoT applications," 2018.

[14] J. Finnegan and S. Brown, "A comparative survey of LPWA networking," arXiv preprint arXiv:1802.04222, 2018.

[15] M. Usama et al., "Unsupervised machine learning for networking: Techniques, applications and research challenges," *IEEE Access*, vol. 7, pp. 65579–65615, 2019.

[16] S. Ayoubi et al., "Machine learning for cognitive network management," *IEEE Communications Magazine*, vol. 56, no. 1, pp. 158–165, 2018.

[17] Z. M. Fadlullah et al., "State-of-the-art deep learning: Evolving machine intelligence toward tomorrow's intelligent network traffic control systems," *IEEE Communications Surveys & Tutorials*, vol. 19, no. 4, pp. 2432–2455, 2017.

[18] M. Wang, Y. Cui, X. Wang, S. Xiao, and J. Jiang, "Machine learning for networking: Workflow, advances and opportunities," *IEEE Network*, vol. 32, no. 2, pp. 92–99, 2017.

[19] C. A. Hammerschmidt, S. Garcia, S. Verwer, and R. State, "Reliable machine learning for networking: Key issues and approaches," In *2017 IEEE 42nd Conference on Local Computer Networks (LCN)*, 2017, pp. 167–170: IEEE.

[20] P. Casas, J. Vanerio, and K. Fukuda, "GML learning, a generic machine learning model for network measurements analysis," In *2017 13th International Conference on Network and Service Management (CNSM)*, 2017, pp. 1–9: IEEE.

[21] C. Modi, D. Patel, B. Borisaniya, H. Patel, A. Patel, and M. Rajarajan, "A survey of intrusion detection techniques in cloud," *Journal of Network and Computer Applications*, vol. 36, no. 1, pp. 42–57, 2013.

[22] K. E. Nolan, W. Guibene, and M. Y. Kelly, "An evaluation of low power wide area network technologies for the Internet of Things," In *2016 International Wireless Communications and Mobile Computing Conference (IWCMC)*, 2016, pp. 439–444: IEEE.

[23] U. Raza, P. Kulkarni, and M. Sooriyabandara, "Low power wide area networks: An overview," *IEEE Communications Surveys & Tutorials*, vol. 19, no. 2, pp. 855–873, 2017.

[24] N. Poursafar, M. E. E. Alahi, and S. Mukhopadhyay, "Long-range wireless technologies for IoT applications: A review," In *2017 Eleventh International Conference on Sensing Technology (ICST)*, 2017, pp. 1–6: IEEE.

[25] J. de Carvalho Silva, J. J. Rodrigues, A. M. Alberti, P. Solic, and A. L. Aquino, "LoRaWAN-A low power WAN protocol for Internet of Things: A review and opportunities," In *2017 2nd International Multidisciplinary Conference on Computer and Energy Science (SpliTech)*, 2017, pp. 1–6: IEEE.

[26] K. Mekki, E. Bajic, F. Chaxel, and F. Meyer, "A comparative study of LPWAN technologies for large-scale IoT deployment," *ICT Express*, vol. 5, no. 1, pp. 1–7, 2019.

[27] M. Aazam, S. Zeadally, and K. A. Harras, "Offloading in fog computing for IoT: Review, enabling technologies, and research opportunities," *Future Generation Computer Systems*, vol. 87, pp. 278–289, 2018.

[28] S. Al-Sarawi, M. Anbar, K. Alieyan, and M. Alzubaidi, "Internet of Things (IoT) communication protocols," In *2017 8th International Conference on Information Technology (ICIT)*, 2017, pp. 685–690: IEEE.

[29] K. Zheng, S. Zhao, Z. Yang, X. Xiong, and W. Xiang, "Design and implementation of LPWA-based air quality monitoring system," *IEEE Access*, vol. 4, pp. 3238–3245, 2016.

[30] W. Guibene, J. Nowack, N. Chalikias, K. Fitzgibbon, M. Kelly, and D. Prendergast, "Evaluation of LPWAN technologies for smart cities: River monitoring use-case," In *2017 IEEE Wireless Communications and Networking Conference Workshops (WCNCW)*, 2017, pp. 1–5: IEEE.

[31] W. Guibene, K. E. Nolan, and M. Y. Kelly, "Survey on clean slate cellular-iot standard proposals," In *2015 IEEE International Conference on Computer and Information Technology; Ubiquitous Computing and Communications; Dependable, Autonomic and Secure Computing; Pervasive Intelligence and Computing*, 2015, pp. 1596–1599: IEEE.

[32] F. Adelantado, X. Vilajosana, P. Tuset-Peiro, B. Martinez, J. Melia-Segui, and T. Watteyne, "Understanding the limits of LoRaWAN," *IEEE Communications Magazine*, vol. 55, no. 9, pp. 34–40, 2017.

[33] L. Gregora, L. Vojtech, and M. Neruda, "Indoor signal propagation of LoRa technology," In *2016 17th International Conference on Mechatronics-Mechatronika (ME)*, 2016, pp. 1–4: IEEE.

[34] B. Reynders, W. Meert, and S. Pollin, "Power and spreading factor control in low power wide area networks," In *2017 IEEE International Conference on Communications (ICC)*, 2017, pp. 1–6: IEEE.

[35] N. Mangalvedhe, R. Ratasuk, and A. Ghosh, "NB-IoT deployment study for low power wide area cellular IoT," In *2016 IEEE 27th Annual International Symposium on Personal, Indoor, and Mobile Radio Communications (PIMRC)*, 2016, pp. 1–6: IEEE.

[36] J. Petajajarvi, K. Mikhaylov, A. Roivainen, T. Hanninen, and M. Pettissalo, "On the coverage of LPWANs: Range evaluation and channel attenuation model for LoRa technology," In *2015 14th International Conference on ITS Telecommunications (ITST)*, 2015, pp. 55–59: IEEE.

[37] M. Aref and A. Sikora, "Free space range measurements with Semtech LoRa(tm) technology," In *2014 2nd International Symposium on Wireless Systems within the Conferences on Intelligent Data Acquisition and Advanced Computing Systems*, 2014, pp. 19–23: IEEE.

[38] M. Lauridsen, I. Z. Kovács, P. Mogensen, M. Sorensen, and S. Holst, "Coverage and capacity analysis of LTE-M and NB-IoT in a rural area," In *2016 IEEE 84th Vehicular Technology Conference (VTC-Fall)*, 2016, pp. 1–5: IEEE.

[39] F. Mattern and C. Floerkemeier, "From the internet of computers to the internet of things," From Active Data Management to Event-Based Systems and More: Papers in honor of Alejandro Buchmann on the Occasion of His 60th Birthday, pp. 242–259, 2010.

[40] S. Tanwar, J. Garg, M. Gupta, and A. Rana, "Opportunities and challenges in machine learning with IoT," *Machine Learning Paradigm for Internet of Things Applications*, pp. 209–227, 2022.

[41] B. Guo, D. Zhang, Z. Yu, Y. Liang, Z. Wang, and X. Zhou, "From the internet of things to embedded intelligence," *World Wide Web*, vol. 16, pp. 399–420, 2013.

[42] B. Guo, D. Zhang, and Z. Wang, "Living with internet of things: The emergence of embedded intelligence," In *2011 International Conference on Internet of Things and 4th International Conference on Cyber, Physical and Social Computing*, 2011, pp. 297–304: IEEE.

[43] D. A. Keim, T. Munzner, F. Rossi, and M. Verleysen, "Bridging information visualization with machine learning (Dagstuhl Seminar 15101)," *Dagstuhl Reports*, vol. 5, no. 3, 2015. Schloss Dagstuhl-Leibniz-Zentrum fuer Informatik.

[44] J. Bzai et al., "Machine learning-enabled Internet of Things (IoT): Data, applications, and industry perspective," *Electronics*, vol. 11, no. 17, p. 2676, 2022.

[45] Accenture Strategy, "The growth game-changer: How the industrial internet of things can drive progress and prosperity. 2015," *Acesso em*, vol. 17, 2017.

[46] A. Kearney, "The Internet of Things: A new path to European prosperity," *AT Kearney*, 2016.

[47] A. Z. C. P. D. Georgakopoulos, A. Zaslavsky, and C. Perera, "Sensing as a service and big data," In *Proceedings of the International Conference on Advances in Cloud Computing (ACC'12)*, 2012.

[48] D. E. O'Leary, "Big Data, the 'Internet of Things' and the 'Internet of Signs'," *Intelligent Systems in Accounting, Finance and Management*, vol. 20, no. 1, pp. 53–65, 2013.

[49] Y. Sun, H. Song, A. J. Jara, and R. Bie, "Internet of things and big data analytics for smart and connected communities," *IEEE Access*, vol. 4, pp. 766–773, 2016.

[50] M. Lindstrom, *Small Data: The Tiny Clues That Uncover Huge Trends*. St. Martin's Press, 2016.

[51] F. Chen, P. Deng, J. Wan, D. Zhang, A. V. Vasilakos, and X. Rong, "Data mining for the internet of things: Literature review and challenges," *International Journal of Distributed Sensor Networks*, vol. 11, no. 8, p. 431047, 2015.

[52] Y. Qin, Q. Z. Sheng, N. J. Falkner, S. Dustdar, H. Wang, and A. V. Vasilakos, "When things matter: A survey on data-centric internet of things," *Journal of Network and Computer Applications*, vol. 64, pp. 137–153, 2016.

[53] L. Zhang, D. Jeong, and S. Lee, "Data quality management in the internet of things," *Sensors*, vol. 21, no. 17, p. 5834, 2021.

[54] C. C. Aggarwal, N. Ashish, and A. Sheth, "The internet of things: A survey from the data-centric perspective," *Managing and Mining Sensor Data*, C. C. Aggarwal, ed., Springer, New York, pp. 383–428, 2013.

[55] M. Abu-Elkheir, M. Hayajneh, and N. A. Ali, "Data management for the internet of things: Design primitives and solution," *Sensors*, vol. 13, no. 11, pp. 15582–15612, 2013.

[56] S. Cleland, "Google's 'Infringenovation' Secrets," https://www.forbes.com/sites/scottcleland/2011/10/03/googles-infringenova-tion-secrets/?sh=1c45fc2c30a6, 2011.

[57] P. Domingos, "A few useful things to know about machine learning," *Communications of the ACM*, vol. 55, no. 10, pp. 78–87, 2012.

[58] A. Halevy, P. Norvig, and F. Pereira, "The unreasonable effectiveness of data," *IEEE Intelligent Systems*, vol. 24, no. 2, pp. 8–12, 2009.

[59] V. Sessions and M. Valtorta, "The effects of data quality on machine learning algorithms," *ICIQ*, vol. 6, pp. 485–498, 2006.

[60] C. Cortes, L. D. Jackel, and W.-P. Chiang, "Limits on learning machine accuracy imposed by data quality," *Advances in Neural Information Processing Systems*, vol. 7, 1994.

[61] M. Ambasna-Jones, "Will automation and the internet of things lead to mass unemployment," *The Guardian*, 2015.

[62] H. Stewart, "Robot revolution: Rise of 'thinking' machines could exacerbate inequality," *The Guardian*, vol. 5, p. 2015, 2015.

[63] M. Banko and E. Brill, "Mitigating the paucity-of-data problem: Exploring the effect of training corpus size on classifier performance for natural language processing," In *Proceedings of the First International Conference on Human Language Technology Research*, 2001.

[64] V. Moreno-Cano, F. Terroso-Saenz, and A. F. Skarmeta-Gomez, "Big data for IoT services in smart cities," In *2015 IEEE 2nd World Forum on Internet of Things (WF-IoT)*, 2015, pp. 418–423: IEEE.

[65] U. S. Shanthamallu, A. Spanias, C. Tepedelenlioglu, and M. Stanley, "A brief survey of machine learning methods and their sensor and IoT applications," In *2017 8th International Conference on Information, Intelligence, Systems & Applications (IISA)*, 2017, pp. 1–8: IEEE.

[66] K. Sharma and R. Nandal, "A literature study on machine learning fusion with IOT," In *2019 3rd International Conference on Trends in Electronics and Informatics (ICOEI)*, 2019, pp. 1440–1445: IEEE.

[67] P. K. Donta, S. N. Srirama, T. Amgoth, and C. S. R. Annavarapu, "Survey on recent advances in IoT application layer protocols and machine learning scope for research directions," *Digital Communications and Networks*, vol. 8, no. 5, pp. 727–744, 2022.

[68] C. Monga, D. Gupta, D. Prasad, S. Juneja, G. Muhammad, and Z. Ali, "Sustainable network by enhancing attribute-based selection mechanism using Lagrange interpolation," *Sustainability*, vol. 14, no. 10, p. 6082, 2022.

[69] P. Singh et al., "A fog-cluster based load-balancing technique," *Sustainability*, vol. 14, no. 13, p. 7961, 2022.

[70] S. Kanwal, J. Rashid, J. Kim, S. Juneja, G. Dhiman, and A. Hussain, "Mitigating the coexistence technique in wireless body area networks by using superframe interleaving," *IETE Journal of Research*, pp. 1–15, 2022.

[71] L. Atzori, A. Iera, and G. Morabito, "The internet of things: A survey," *Computer Networks*, vol. 54, no. 15, pp. 2787–2805, 2010.

[72] A. Whitmore, A. Agarwal, and L. Da Xu, "The internet of things: A survey of topics and trends," *Information Systems Frontiers*, vol. 17, pp. 261–274, 2015.

[73] L. Da Xu, W. He, and S. Li, "Internet of things in industries: A survey," *IEEE Transactions on Industrial Informatics*, vol. 10, no. 4, pp. 2233–2243, 2014.

[74] M. N. Kamel Boulos and N. M. Al-Shorbaji, "On the Internet of Things, smart cities and the WHO healthy cities," *International Journal of Health Geographics*, vol. 13, pp. 1–6, 2014.

[75] P. High, "The top five smart cities in the world," *Forbes*, 2015.

[76] V. K. Garg and S. Sharma, "Role of IoT and machine learning in smart grid," *Emergent Converging Technologies and Biomedical Systems: Select Proceedings of ETBS 2021*: Springer, 2022, pp. 701–715.

[77] R. E. Bryant, R. H. Katz, C. Hensel, and E. P. Gianchandani, "From data to knowledge to action: Enabling the smart grid," *arXiv preprint arXiv:2008.00055*, 2020.

[78] A. S. Ali and S. Azad, "Demand forecasting in smart grid," *Smart Grids: Opportunities, Developments, and Trends*, pp. 135–150, 2013.

[79] M. E. Khodayar and H. Wu, "Demand forecasting in the smart grid paradigm: Features and challenges," *The Electricity Journal*, vol. 28, no. 6, pp. 51–62, 2015.

[80] A. Kotillova, I. Koprinska, and M. Rana, "Statistical and machine learning methods for electricity demand prediction," In *Neural Information Processing: 19th International Conference, ICONIP 2012, Doha, Qatar, November 12-15, 2012, Proceedings, Part II 19*, 2012, pp. 535–542: Springer, Berlin Heidelberg.

[81] A. Khotanzad, H. Elragal, and T.-L. Lu, "Combination of artificial neural-network forecasters for prediction of natural gas consumption," *IEEE Transactions on Neural Networks*, vol. 11, no. 2, pp. 464–473, 2000.

[82] J.-m. Wang and X.-h. Liang, "The forecast of energy demand on artificial neural network," In *2009 International Conference on Artificial Intelligence and Computational Intelligence*, 2009, vol. 3, pp. 31–35: IEEE.

[83] B. Chen, L. Zhao, and J. H. Lu, "Wind power forecast using RBF network and culture algorithm," In *2009 International Conference on Sustainable Power Generation and Supply*, 2009, pp. 1–6: IEEE.

[84] W.-Y. Chang, "Short-term load forecasting using radial basis function neural network," *Journal of Computer and Communications*, vol. 3, no. 11, p. 40, 2015.

[85] P. Musilek, E. Pelikán, T. Brabec, and M. Simunek, "Recurrent neural network based gating for natural gas load prediction system," In *The 2006 IEEE International Joint Conference on Neural Network Proceedings*, 2006, pp. 3736–3741: IEEE.

[86] M. A. El-Sharkawi and R. Marks, "Applications of neural networks to power systems," *Seattle (WA)*, vol. 231, p. 236, 1991.

[87] J. Kumaran and G. Ravi, "Long-term sector-wise electrical energy forecasting using artificial neural network and biogeography-based optimization," *Electric Power Components and Systems*, vol. 43, no. 11, pp. 1225–1235, 2015.

[88] Z. Aung, M. Toukhy, J. Williams, A. Sanchez, and S. Herrero, "Towards accurate electricity load forecasting in smart grids," In *Proceedings of DBKDA, International Academy, Research, and Industry Association (IARIA)*, USA, 2012, pp. 51–57.

[89] D. Masri, H. Zeineldin, and W. L. Woon, "Electricity price and demand forecasting under smart grid environment," In *2015 IEEE 15th International Conference on Environment and Electrical Engineering (EEEIC)*, 2015, pp. 1956–1960: IEEE.

[90] Y. Sun, U. Braga-Neto, and E. R. Dougherty, "Impact of missing value imputation on classification for DNA microarray gene expression data-a model-based study," *EURASIP Journal on Bioinformatics and Systems Biology*, vol. 2009, pp. 1–17, 2010.

[91] E. Mocanu, P. H. Nguyen, M. Gibescu, E. M. Larsen, and P. Pinson, "Demand forecasting at low aggregation levels using factored conditional restricted boltzmann machine," In *2016 Power Systems Computation Conference (PSCC)*, 2016, pp. 1–7: IEEE.

[92] E. Hossain, I. Khan, F. Un-Noor, S. S. Sikander, and M. S. H. Sunny, "Application of big data and machine learning in smart grid, and associated security concerns: A review," *IEEE Access*, vol. 7, pp. 13960–13988, 2019.

[93] D. J. Cook, M. Youngblood, and S. K. Das, "A multi-agent approach to controlling a smart environment," *Designing Smart Homes: The Role of Artificial Intelligence*, Juan Carlos Augusto, Chris D. Nugent, eds., Springer Berlin, Heidelberg, pp. 165–182, 2006.

[94] M. H. Kabir, M. R. Hoque, H. Seo, and S.-H. Yang, "Machine learning based adaptive context-aware system for smart home environment," *International Journal of Smart Home*, vol. 9, no. 11, pp. 55–62, 2015.

[95] A. Dixit and A. Naik, "Use of prediction algorithms in smart homes," *International Journal of Machine Learning and Computing*, vol. 4, no. 2, p. 157, 2014.

[96] O. Taiwo, A. E. Ezugwu, O. N. Oyelade, and M. S. Almutairi, "Enhanced intelligent smart home control and security system based on deep learning model," *Wireless Communications and Mobile Computing*, vol. 2022, pp. 1–22, 2022.

[97] S.-H. Lee and C.-S. Yang, "An intelligent home access control system using deep neural network," In *2017 IEEE International Conference on Consumer Electronics-Taiwan (ICCE-TW)*, 2017, pp. 281–282: IEEE.

[98] V. Bellandi, P. Ceravolo, E. Damiani, and S. Siccardi, "Smart Healthcare, IoT and Machine Learning: A Complete Survey," *Handbook of Artificial Intelligence in Healthcare: Vol 2: Practicalities and Prospects*, Chee-Peng Lim, Yen-Wei Chen, Ashlesha Vaidya, Charu Mahorkar, Lakhmi C. Jain, eds., Springer, Cham, pp. 307–330, 2022.

[99] T. Chatzinikolaou, E. Vogiatzi, A. Kousis, and C. Tjortjis, "Smart health-care support using data mining and machine learning," *IoT and WSN based Smart Cities: A Machine Learning* Perspective: Springer, pp. 27–48, 2022.

[100] O. Boursalie, R. Samavi, and T. E. Doyle, "M4CVD: Mobile machine learning model for monitoring cardiovascular disease," *Procedia Computer Science*, vol. 63, pp. 384–391, 2015.

[101] N. Vyas, J. Farringdon, D. Andre, and J. I. Stivoric, "Machine learning and sensor fusion for estimating continuous energy expenditure," *AI Magazine*, vol. 33, no. 2, pp. 55–55, 2012.

[102] S. Smith, "Ways the IBM Watson is changing health care, from diagnosing disease to treating it," ed.https://www.medicaldaily.com/5-ways-ibm-watson-changing-health-care-diagnosing-disease-treating-it-364394

[103] B. Liu and L. Zhang, "A survey of opinion mining and sentiment analysis," *Mining Text Data*, C. C. Aggarwal, C. Zhai, eds., Springer, New York, pp. 415–463, 2012.

[104] R. Bouchlaghem, A. Elkhelifi, and R. Faiz, "A machine learning approach for classifying sentiments in Arabic tweets," In *Proceedings of the 6th International Conference on Web Intelligence, Mining and Semantics*, 2016, pp. 1–6.

[105] A. Akbar, F. Carrez, K. Moessner, and A. Zoha, "Predicting complex events for pro-active IoT applications," In *2015 IEEE 2nd World Forum on Internet of Things (WF-IoT)*, 2015, pp. 327–332: IEEE.

[106] P. P. Ray, M. Mukherjee, and L. Shu, "Internet of things for disaster management: State-of-the-art and prospects," *IEEE Access*, vol. 5, pp. 18818–18835, 2017.

[107] N. Nesa, T. Ghosh, and I. Banerjee, "Outlier detection in sensed data using statistical learning models for IoT," In *2018 IEEE Wireless Communications and Networking Conference (WCNC)*, 2018, pp. 1–6: IEEE.

[108] M. Salehi and L. Rashidi, "A survey on anomaly detection in evolving data: [with application to forest fire risk prediction]," *ACM SIGKDD Explorations Newsletter*, vol. 20, no. 1, pp. 13–23, 2018.

[109] G. Pughazhendhi, A. Raja, P. Ramalingam, and D. K. Elumalai, "Earthosys-tsunami prediction and warning system using machine learning and IoT," In *Proceedings of International Conference on Computational Intelligence and Data Engineering: Proceedings of ICCIDE 2018*, 2019, pp. 103–113: Springer.

[110] Y. Andaloussi, M. D. El Ouadghiri, Y. Maurel, J.-M. Bonnin, and H. Chaoui, "Access control in IoT environments: Feasible scenarios," *Procedia Computer Science*, vol. 130, pp. 1031–1036, 2018.

[111] L. Xiao, X. Wan, X. Lu, Y. Zhang, and D. Wu, "IoT security techniques based on machine learning: How do IoT devices use AI to enhance security?," *IEEE Signal Processing Magazine*, vol. 35, no. 5, pp. 41–49, 2018.

[112] F. Hussain, R. Hussain, S. A. Hassan, and E. Hossain, "Machine learning in IoT security: Current solutions and future challenges," *IEEE Communications Surveys & Tutorials*, vol. 22, no. 3, pp. 1686–1721, 2020.

[113] M. Hasan, M. M. Islam, M. I. I. Zarif, and M. Hashem, "Attack and anomaly detection in IoT sensors in IoT sites using machine learning approaches," *Internet of Things*, vol. 7, p. 100059, 2019.

[114] W. H. Khalifa, M. I. Roushdy, and A.-B. M. Salem, "Machine learning techniques for intelligent access control," *New Approaches in Intelligent Image Analysis: Techniques, Methodologies and Applications*, Roumen Kountchev, Kazumi Nakamatsu, eds., Springer, Cham, pp. 331–351, 2016.

[115] B. Marr, "What is industry 4.0? Here's a super easy explanation for anyone. [online]," *Forbes*, 2018.

[116] R. Barga, V. Fontama, W. H. Tok, and L. Cabrera-Cordon, *Predictive analytics with Microsoft Azure Machine Learning*: Springer, 2015.

[117] S. Higginbotham, "IBM is bringing in Watson to conquer the Internet of Things," ed: Fortune, 2015.

[118] B. Kepes, "Splunk ups the machine-learning ante," *Computerworld*, July, 2017.

[119] T. S. Brisimi, C. G. Cassandras, C. Osgood, I. C. Paschalidis, and Y. Zhang, "Sensing and classifying roadway obstacles in smart cities: The street bump system," *IEEE Access*, vol. 4, pp. 1301–1312, 2016.

[120] E. Migabo, K. Djouani, A. Kurien, and T. Olwal, "A comparative survey study on LPWA networks: LoRa and NB-IoT," In *Proceedings of the Future Technologies Conference (FTC)*, Vancouver, BC, Canada, 2017, pp. 29–30.

[121] "AN1200.22: LoRa Modulation Basics", Semtech, 2015.

[122] M. I. Hossain, L. Lin, and J. Markendahl, "A comparative study of IoT-communication systems cost structure: Initial findings of radio access networks cost," In *2018 11th CMI International Conference: Prospects and Challenges Towards Developing a Digital Economy within the EU*, 2018, pp. 49–55: IEEE. https://ieeexplore.ieee.org/document/8624853

[123] A. Šljivo et al., "Performance evaluation of IEEE 802.11 ah networks with high-throughput bidirectional traffic," *Sensors*, vol. 18, no. 2, p. 325, 2018.

[124] K. Mikhaylov, J. Petaejaejaervi, and T. Haenninen, "Analysis of capacity and scalability of the LoRa low power wide area network technology," In *European Wireless 2016; 22th European Wireless Conference*, 2016, pp. 1–6: VDE.

[125] F. Muteba, K. Djouani, and T. Olwal, "A comparative survey study on LPWA IoT technologies: Design, considerations, challenges and solutions," *Procedia Computer Science*, vol. 155, pp. 636–641, 2019.

[126] S. Nižetić, P. Šolić, D. L.-d.-I. González-De, and L. Patrono, "Internet of Things (IoT): Opportunities, issues and challenges towards a smart and sustainable future," *Journal of Cleaner Production*, vol. 274, p. 122877, 2020.

[127] K. M. Sadique, R. Rahmani, and P. Johannesson, "Towards security on internet of things: Applications and challenges in technology," *Procedia Computer Science*, vol. 141, pp. 199–206, 2018.

[128] Y. Rahul, Z. Weizhe, K. Neeraj, and K. Omprakash, *Recent trends and challenges in fog/edge computing for Internet of Things* (accessed on 2 June 2022). https://www.hindawi.com/journals/wcmc/si/975923/

[129] A. A. Abi Sen, F. A. Eassa, K. Jambi, and M. Yamin, "Preserving privacy in internet of things: A survey," *International Journal of Information Technology*, vol. 10, pp. 189–200, 2018.

[130] R. Falcone and A. Sapienza, "On the users' acceptance of IoT systems: A theoretical approach," *Information*, vol. 9, no. 3, p. 53, 2018.

[131] F. Beştepe and S. Ö. Yildirim, "Acceptance of IoT-based and sustainability-oriented smart city services: A mixed methods study," *Sustainable Cities and Society*, vol. 80, p. 103794, 2022.

[132] Y. Patil, *6 Key challenges to consider for successful IoT implementation* (accessed on 2 June 2022). https://www.saviantconsulting.com/blog/iot-implementation-challenges-enterprises.aspx

[133] Bernard Marr, *The 5 biggest Internet of Things (IoT) trends in 2021* (accessed on 2 June 2022). https://www.forbes.com/sites/bernardmarr/2020/10/26/the-5-biggest-internet-of-things-iot-trends-in-2021-everyone-must-get-ready-for-now/?sh=5088dcbb41fd

[134] S. R. Chohan and G. Hu, "Success factors influencing citizens' adoption of IoT service orchestration for public value creation in smart government," *IEEE Access*, vol. 8, pp. 208427–208448, 2020.

Chapter 5

Ad Hoc aided cellular networks and LPWAN applications

A review

Abdul Qahar Shahzad
Quaid-i-Azam University

Muhammad Allah Rakha
FAST National University of Computer and Emerging Sciences

Ifrah Komal
Riphah international University

5.1 INTRODUCTION

Mobile-based communication is the optimal source which can support humans economically in every walk of life [1–3]. In the conventional internet approach, the wire line connectivity is utilized [4]. As the world has changed a lot, therefore next generation computing is introduced which is known as 5G [5,6]. However, the previous version mainly comprised of radio-interfaces. However, 5G is used to offer an inclusive approach to integrate wireless and wired-based networks [7]. In addition, new structural transformation in network directly exists in radio-technologies, interfaces, antennas and frequency spectrum. Quality of service metrics is used to improve several applications, which include multi-access edge computing and software-defined networks [8]. As mobile and wireless communication networks have made incredible progress in recent years, 5G growth improves high-tech and mutual interactions [9]. Researchers are continuously working to enhance the capabilities among people and high technology where interactions can be made possible easily [10]. Actor network theory plays an important role in defining applications related to 5G cellular networks [11]. Technological experts trying to improve the performance of radio spectrum-based networks. Fifth-generation-based technology has influenced to advance existing wireless approaches like WLAN approach to Wi-Fi. Accessible WLAN tools are very less expensive but have limited area for coverage. WLAN must use timing protocols to improve the quality of service parameters [12,13]. Therefore, 5G ensures to cover a larger area in comparison with existing technology. ISM-Band can be used to provide dedicated frequencies with a frequency range from 2.4 to 5 GHz. In

DOI: 10.1201/9781003426974-5

accordance, 5G frequencies reduce jamming-based signals [14]. Nowadays, approximately 70% of mobile phone users have Bluetooth which covers limited area [15,16]. While radio-based communication technology uses less energy, the usage of power in WLAN consumes ten times more energy in comparison with Bluetooth [17]. Also, using the mentioned technology, device-to-device connection will have a restricted range. Therefore, 5G protocols operate in long-range communication with a high data exchange rate.

Routing protocols in wireless networks [18] is a trending topic for discussion. Most of the practitioners use DSDV and AODV routing for signal node communication. These routing protocols maintain routing table to avoid loop or redundancy. However, the AODV routing protocol is designed to reshape the paths on special request. AODV routing enhances association and reduces load on every node [19]. Figure 5.1 represents routing protocols and the concept of smart city.

However, the major contribution of the review article includes:

- Wireless communication technologies like 1G to 5G.
- Routing protocols in different classes.
- Brief survey about routing protocols in various fields of study.
- Connectivity issues are discussed thoroughly.
- Applications of Low Power Wide Area Networks

5.2 LITERATURE REVIEW

Wireless communication technologies can be used to have a novel architecture which improves connectivity [20,21]. FDMA and TDMA require

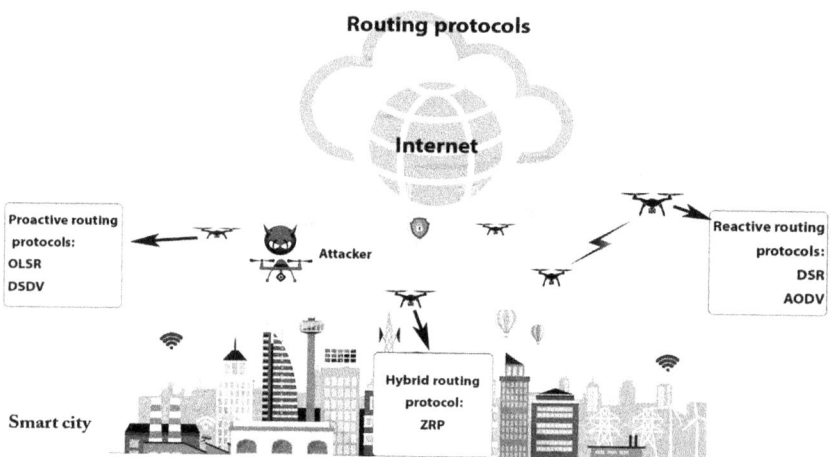

Figure 5.1 Smart cities and routing protocols.

connectivity between stations [22,23]. However, mobile communication also provides the same technique between nodes. First-generation networks cannot be practically used in modern concepts. Therefore, the analog phone comes under the umbrella of 1G cellular network. Due to the arrival of 2G, communication channels were upgraded in the history of mankind. GSM approach was the optimal approach but still not have an Internet option. Technology experts introduced GPRS service especially for browsing in GSM [24]. The cellular phones are converted into digital system which used to have CDMA and GSM version of 2G and 2.5G [25,26]. Due to the function succession of GPRS, a novel approach to EGPRS is initiated which is much more reliable [27]. Then, the innovation of 3G has given speedy features in internet connectivity. Also, after sometimes 3.5G was introduced with high bandwidth on mobile devices like 7.2 megabytes per second [28,29]. Smartphones are linked to personal computers to improve networking problem [30]. Due to the advancement in mobile communication, 4G model is formulated which later refers to novel contributions. Through the innovation of OFDM, which refers to 5G use, browsing speeds of several hundred megabytes per second [31,32]. Moreover, effective communication with high-speed internet has a lot of location constraints [33]. Cellular networks with advanced features can be utilized in IoTs, machines, and other accessories that can interconnect everything together [34]. Wireless technologies are used to provide more coverage to applications related to MANETs, VANETs, and flying ad hoc networks. Therefore, routing protocols play a very important role between nodes. A novel energy-efficient routing protocol called E-AntHocNet is introduced in the area of FANETs. Quality of service parameters such as packet drop rate, throughput, and end-to-end delay are utilized to check the performance of FANETs [35]. Apart from that, machine learning techniques like decision tree are used to improve the signal strength from base station to aerial vehicles [36]. Table 5.1 describes 5G technology usage and applications.

Table 5.1 5G usage and real-time applications

References	Areas where 5G is utilized	5G real-time application	Description
[37]	Medical	United Nations healthcare applications	Through strong connectivity among IoT, ML, AI, and big data it will be possible to reach patients on time
[38]	Internet of things (IoT)	Wireless Sensor Networks (WSNs) energy efficiency	MIMO-HC (multi-input and multi-output hybrid clustering) along with IoT is utilized in 5G environment to improve WSNs efficiency in term of energy consumption

(Continued)

Table 5.1 (Continued) 5G usage and real-time applications

References	Areas where 5G is utilized	5G real-time application	Description
[39]	MANETs	Routing protocols	Radio-system (data transmission route) is longer lasting and energy efficient in MANETs. Through this approach overall performance is enhanced
[40]	Cellular Network	DAWN (Dynamic Adhoc Wireless Network)	QoS is improved and cost is reduced for mobile consumer. At same time through radio and modulation, consumer will operate without interruption
[41]	Seismology	Seismic wave acquisition and interpretation	Seismic Data Acquisition System along with cloud-technology and IoT operate under 5g for rapid seismological interpretation to attain high accuracy
[42]	UAVs	Routing protocols	Through 5G UAVs communication is enhanced due to high-rate transmission
[43]	Warfare	Stealth communication	Confidential information's are shared without low probability of detection or interference
[44]	Smart City Design	Waste recycling protocols	5G assist in constructing and demolishing waste recycling through autonomous transport, imagery, and smart-road able to charge wireless electronic trucks
[45]	NB-IoT	Communication protocols	MF-DR (Multi-Factor Forwarding Dynamic Routing) is utilized to reduce energy consumption and networking traffic of WSNs
[46]	Earthquakes	Early warning system	Pre-Seismic warning message is delivered to buildings or densely populated area in order to evacuate. 5G and SHM system is utilized in this regard

5.3 ROUTING PROTOCOLS

Every new emerging area uses existing routing protocols of mobile ad hoc networks. Therefore, routing protocols are divided into three main types, which include proactive, reactive, and hybrid. Figure 5.2 represents routing protocols which can be utilized in different areas [47].

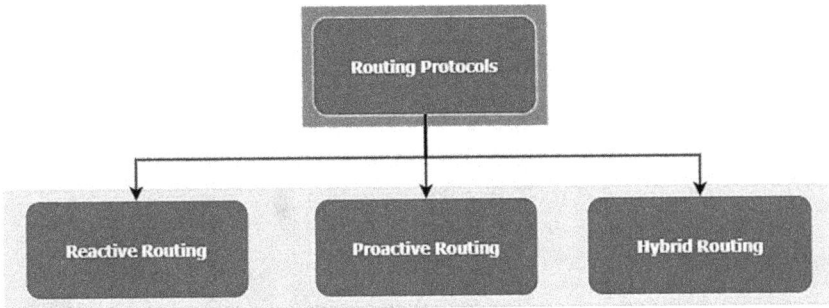

Figure 5.2 Types of routing protocols.

5.3.1 Reactive protocols

In reactive protocols, the routes need to be created when communication starts from source to destination. These routing techniques are called table-driven routing strategies. DSR and AODV are considered reactive routing protocols [48–52].

5.3.2 Proactive protocols

Earlier, every protocol was developed especially for mobile ad hoc networks where nodes were usually static. In the concept of VANETs and FANETs, the nodes are dynamic in comparison with MANETS. As the topological structure normally changes with the passage of time, the routing tables are updated as well. OLSR and DSDV are the best examples for proactive protocols [53–56].

5.3.3 Hybrid protocols

Hybrid protocols are used to have reactive and proactive features. Zone routing protocol (ZRP) is a hybrid protocol which has the capability of clustering and dividing the topology into zones [57]. Table 5.2 shows the related information about routing protocols. Different areas of study are evaluated to check the dynamics of MANETs, VANETs, and FANETs.

5.4 SURVEY STUDY ON WIRELESS COMMUNICATION TECHNOLOGIES

This study provides useful information starting from first generation to 5G [69], which consists of access technologies, data rate, frequency band, bandwidth, forward error correction, switching, and real-time applications. Table 5.3 describes the detailed study about wireless communication technologies.

Table 5.2 A survey on routing protocols deployed in various areas

Reference	Area of study	Routing protocols	Mobility models	Algorithm	Tool	Applications	Description
[58]	FANET	OLSR, AODV, DSR, GRP	RWPM, MGM, SCRM, PRS	TORA algorithm	OPNET 17.5	Crime tracking, Surveillance of city roads	The performance effect of node mobility is higher than the performance effect of node speed
[59]	MANETs	AODV DSR, and DSDV	Random waypoint mobility model	Energy efficient routing algorithm	FEEAODV	Pandemic medical aid in remote locations, wartime military coordination, and other civil research applications	FEEAODV performance output achieves better than AODV performance
[60]	FANET	QMR and Q-Noise+	Waypoint mobility model	Q-Learning algorithm	WSNET	Video streaming	The Q-FANET output is somewhat higher than QMR with the packet delivery ratio indicating a better 0.81% and 1.26% performance for all nodes of the scenario

(Continued)

Table 5.2 (Continued) A survey on routing protocols deployed in various areas

Reference	Area of study	Routing protocols	Mobility models	Algorithm	Tool	Applications	Description
[61]	FANET	AODV, OLSR	RWPM, GMM, and SRCM Gauss–Markov mobility model	TORA	NS3	Tactical and wireless networks, firefighting, search, rescue teams, thermal detection are crucial for the detection of COVID 19 patients	This indicates that the Gauss–Markov mobility model delivers the maximum possible outcome with Ad Hoc On-Demand Vectors, while the 2.4 GHz frequency with the lowest retransmission attempts is as much as 5 GHz.
[62]	FANETs	SrFTime and CRP	3-D mobility models	TORA	ns-3	Smart transport systems, management of wildfires and logistic operations	Use ns-3 simulator and introduced 3-D mobility models for current 802.11s standard
[63]	FANET	MDA-AODV and MA-DP-AODV-AHM MDRMA protocol	Random way point mobility model	MDRMA-Routing algorithm	ns-3	Mobile computing areas	With respect to end-to-end delayed, overhead routing and packet delivery ratios, MDRMA protocol operates very well in all simulation classes and beats its end to end delay

(Continued)

Table 5.2 (Continued) A survey on routing protocols deployed in various areas

Reference	Area of study	Routing protocols	Mobility models	Algorithm	Tool	Applications	Description
[64]	FANET	Aerosp. geo-routing, DSDV	BeeAdHoc based mobility model	Bee algorithm	ns-2	User's location (GLONASS/GPS)	In most cases, algorithms based on the colony of bees have demonstrated more effectiveness than conventional FANET routing algorithms
[65]	FANET	Routing 3D Geographic Greedy forwarding, face routing	Random waypoint mobility model	Ricci flow, Mobus transform, geodesic pattern	TOSSIM	Smart cities, smart homes, clever agriculture, smart grid, environment monitoring	Their results were compared with GDSTR-3D and HWE, showing that trace routing is better and stable with localization faults
[66]	UAVs and VANETs	AODV TBRF DSDV, OLSR, GRP	Random waypoint mobility model	Temporally Ordered Routing algorithm (TORA)	ns-3	Military apps, medical applications	Due to its high performance and low delay, AODV routing protocol with UAVs provides a great performance for all UAVs
[67]	UAV networks	OLSR	NIL	Link state algorithm	OPNET	Military and civil applications	The OLSR in FANET's best choice in terms of latency, network loading, transmission and retransmission
[68]	FANET AND VANET	DSDV	Random	TORA	NIL	Tenacity, weather, Military, civilian and Security	SANET's new approach is to demonstrate the progress of the Sea ad hoc network

Table 5.3 Comparative analysis of wireless technologies

Reference	Genera	Access technology	Data rate	Frequency band	Bandwidth	Forward error correction	Switching	Applications
[70]	1G	AMPs, FDMA	2.4 kbps	824–894MHz	30kHz	NA	Circuit	Voice
[66]	2G	TDMA, GSM	9.6 kbps	850/900/1800/1900MHz	200kHz	NA	Circuit	Voice and data
[46]	2.5G	EDGE, GPRS	50 kbps	200 KHz	1.25MHz	Turbo codes	Circuit/Packet	Voice and data
[66]	3G	UMTS, CDMA	384 kbps	800/850/900/1800/1900/2100MHz	2MHz	Turbo codes	Circuit/Packet	Voice+Data+Video calling
[71]	3.5G	EVDO, HSUPA/HSDPA	5–30 Mbps	1.25MHz		Turbo codes	packet	Voice+Data+Video calling
[72]	3.75G	OFDMA / SC-FDMA Fixed WIMAX	100–200 Mbps	3.5GHz and 5.8GHz	10MHz in 5.8GHz	Concatenated codes	packet	Online gaming High definition television
[73]	4G	LTE-A WIMAX, SOFDMA	100–200 Mbps	2.5GHz, and 3.5GHz	7MHz, 5MHz, 10MHz, and 8.75MHz	Turbo codes	packet	Online gaming+High definition Worldwide Interoperability for Television
[74]	5G cellular network architecture	Massive MIMO technology, and Device-to-Device Communication (D2D)	10–50 Gbps	1.8, 2.6GHz and expected 30–300 GHz	60GHz	Low Density Parity Check Codes (LDPC)	Packet	Ultra-High definition video+Virtual reality applications

5.5 APPLICATIONS OF LOW POWER WIDE AREA NETWORKS

LPWAN [75] has attracted the researcher's community by providing possible solutions for IoT-based networks [16]. However, IoT devices are quite vulnerable. So, DoS and DDoS are considered very common attacks in IoT networks. Asrin Abdollahi et al. proposed a novel intrusion detection system [76] to identify security threats [57]. Applications regarding LPWAN are as follows [77].

5.5.1 Short-range M-to-M like zigbee

Zigbee covers a wide range in comparison with traditional approaches.

5.5.2 Bluetooth

Special interest group uses to manage Bluetooth.

5.5.3 Wi-Fi

WIFI is also called 802.11, which is the standard of IEEE and has a very limited range.

5.6 CONCLUSION

The technological advancement in ad hoc networks and IoT needs reliable connectivity between nodes. Therefore, ad hoc networks use wireless communication networks as backbone. Due to that, connectivity issues are mostly resolved. LPWAN usually has low range of communication which includes real-time applications such as ZigBee, Bluetooth, and Wi-Fi. However, routing protocols help to efficiently communicate within nodes. This is a review article that focuses mainly on MANETs, VANETs, and FANETs routing protocols, wireless communication technologies, and short-range applications related to LPWAN. In addition, this study will help researchers, practitioners, engineers, and scientists.

REFERENCES

[1] Andela, Cornelie D., et al. "Mechanisms in endocrinology: Cushing's syndrome causes irreversible effects on the human brain: a systematic review of structural and functional magnetic resonance imaging studies." *European Journal of Endocrinology* 173.1 (2015): 1–14.

[2] Gerpott, Torsten J., & Thomas, Sandra. "Empirical research on mobile Internet usage: A meta-analysis of the literature." *Telecommunications Policy* 38.3 (2014): 291–310.

[3] Lambertz, U., et al. "Small RNAs derived from tRNAs and rRNAs are highly enriched in exosomes from both old and new world Leishmania providing evidence for conserved exosomal RNA Packaging." *BMC Genomics* 16.1 (2015): 1–26.

[4] Trevisan, Martino, et al. "Five years at the edge: Watching internet from the isp network." *IEEE/ACM Transactions on Networking* 28.2 (2020): 561–574.

[5] Forge, Simon, & Vu, Khuong. "Forming a 5G strategy for developing countries: A note for policy makers." *Telecommunications Policy* 44.7 (2020): 101975.

[6] Schneir, Juan Rendon, et al. "A business case for 5G mobile broadband in a dense urban area." *Telecommunications Policy* 43.7 (2019): 101813.

[7] Fitzek, Frank HP, Granelli, Fabrizio, & Seeling, Patrick eds. *Computing in Communication Networks: From Theory to Practice.* Academic Press, 2020.

[8] Liu, Qian, et al. "5G development in China: From policy strategy to user-oriented architecture." *Mobile Information Systems* 2017 (2017).

[9] Sharma, P. "Evolution of mobile wireless communication networks-1G to 5G as well as future prospective of next generation communication network." *International Journal of Computer Science and Mobile Computing* 2.8 (2013): 47–53.

[10] Shantharama, Prateek, et al. "LayBack: SDN management of multi-access edge computing (MEC) for network access services and radio resource sharing." *IEEE Access* 6 (2018): 57545–57561.

[11] Porambage, Pawani, et al. "Survey on multi-access edge computing for internet of things realization." *IEEE Communications Surveys & Tutorials* 20.4 (2018): 2961–2991.

[12] López-Pérez, David, et al. "IEEE 802.11 be extremely high throughput: The next generation of Wi-Fi technology beyond 802.11 ax." *IEEE Communications Magazine* 57.9 (2019): 113–119.

[13] Nurchis, Maddalena, & Bellalta, Boris. "Target wake time: Scheduled access in IEEE 802.11 ax WLANs." *IEEE Wireless Communications* 26.2 (2019): 142–150.

[14] Saleem, Rashid, et al. "An FSS based multiband MIMO system incorporating 3D antennas for WLAN/WiMAX/5G cellular and 5G Wi-Fi applications." *IEEE Access* 7 (2019): 144732–144740.

[15] Mohapatra, Selli, & Kanungo, Priyadarshi. "Performance analysis of AODV, DSR, OLSR and DSDV routing protocols using NS2 Simulator." *Procedia Engineering* 30 (2012): 69–76.

[16] Yousef, Jasemian, & Lars, A. N. "Validation of a real-time wireless telemedicine system, using bluetooth protocol and a mobile phone, for remote monitoring patient in medical practice." *European Journal of Medical Research* 10.6 (2005): 254–262.

[17] Abd Rahman, Abdul Hadi, & Zukarnain, Zuriati Ahmad. "Performance comparison of AODV, DSDV and I-DSDV routing protocols in mobile ad hoc networks." *European Journal of Scientific Research* 31.4 (2009): 556–576.

[18] Couldry, Nick, "Actor network theory and media: do they connect and on what terms?", Hepp, Andreas, Krotz, Friedrich, Moores, Shaun and Winter, Carsten, (eds.) Connectivity, Networks and Flows: Conceptualizing Contemporary Communications. Hampton Publishing, Cresskill, NJ, USA, pp. 93-110, 2008.

[19] Nayyar, Anand. "Flying adhoc network (FANETs): Simulation based performance comparison of routing protocols: AODV, DSDV, DSR, OLSR, AOMDV and HWMP." In *2018 International Conference on Advances in Big Data, Computing and Data Communication Systems (icABCD).* IEEE, 2018.

[20] Al Mehairi, Saeed O., Barada, Hassan, & Al Qutayri, Mahmoud. "Integration of technologies for smart home application." In *2007 IEEE/ACS International Conference on Computer Systems and Applications.* IEEE, 2007.

[21] Pering, T., Agarwal, Y., Gupta, R., & Want, R. "CoolSpots: Reducing the power consumption of wireless mobile devices with multiple radio interfaces." In *Proceedings of ACM MobiSys* (pp. 220–232), ACM. 2006.

[22] Jung, Peter, Baier, Paul Walter, & Steil, Andreas. "Advantages of CDMA and spread spectrum techniques over FDMA and TDMA in cellular mobile radio applications." *IEEE Transactions on Vehicular Technology* 42.3 (1993): 357–364.

[23] Rubin, Izhak. "Message delays in FDMA and TDMA communication channels." *IEEE Transactions on Communications* 27.5 (1979): 769–777.

[24] Adachi, Fumiyuki. "Wireless past and future–evolving mobile communications systems." *IEICE Transactions on Fundamentals of Electronics, Communications and Computer Sciences* 84.1 (2001): 55–60.

[25] Grieco, Donald M., & Schilling, Donald L. "The capacity of broadband CDMA overlaying a GSM cellular system." In *Proceedings of IEEE Vehicular Technology Conference (VTC).* IEEE, 1994.

[26] Venkatesan, K.G.S. "Comparison of CDMA & GSM mobile technology." *Middle-East Journal of Scientific Research* 13.12 (2013): 1590–1594.

[27] Rocha-Osorio, C. M., et al. "GPRS/EGPRS standards applied to DTC of a DFIG using fuzzy-PI controllers." *International Journal of Electrical Power & Energy Systems* 93 (2017): 365–373.

[28] Choi, Tae-Il, et al. "Communication system for distribution automation using CDMA." *IEEE Transactions on Power Delivery* 23.2 (2008): 650–656.

[29] Jou, Y. "Developments in third generation (3G) CDMA technology." In *2000 IEEE Sixth International Symposium on Spread Spectrum Techniques and Applications. ISSTA 2000. Proceedings (Cat. No. 00TH8536).* Vol. 2. IEEE, 2000.

[30] Derryberry, R. Thomas, et al. "Transmit diversity in 3G CDMA systems." *IEEE Communications Magazine* 40.4 (2002): 68–75.

[31] Seong, Kibeom, Mohseni, Mehdi, & Cioffi, John M.. "Optimal resource allocation for OFDMA downlink systems." In *2006 IEEE International Symposium on Information Theory.* IEEE, 2006.

[32] Yin, Hujun, & Alamouti, Siavash. "OFDMA: A broadband wireless access technology." *2006 IEEE Sarnoff Symposium.* IEEE, 2006.

[33] Hara, Shinsuke, & Prasad, Ramjee. *Multicarrier techniques for 4G mobile communications.* Artech House, 2003.

[34] Cox, Christopher. *An introduction to LTE: LTE, LTE-advanced, SAE and 4G mobile communications.* John Wiley & Sons, 2012.

[35] Khan, I. U., Qureshi, I. M., Aziz, M. A., Cheema, T. A., & Shah, S. B. H. "Smart IoT control-based nature inspired energy efficient routing protocol for flying ad hoc network (FANET)." *IEEE Access* 8 (2020): 56371–56378. https://doi.org/10.1109/ACCESS.2020.2981531.

[36] Khan, I. U., Alturki, R., Alyamani, H. J., Ikram, M. A., Aziz, M. A., Hoang, V. T., & Cheema, T. A. "RSSI-controlled long-range communication in secured IoT-enabled unmanned aerial vehicles." *Mobile Information Systems* 2021 (2021): 1–11.

[37] Latif, Siddique, et al. "How 5G wireless (and concomitant technologies) will revolutionize healthcare?" *Future Internet* 9.4 (2017): 93.

[38] Baniata, Mohammad, et al. "Energy-efficient hybrid routing protocol for IoT communication systems in 5G and beyond." *Sensors* 21.2 (2021): 537.

[39] Quy, Vu Khanh, Ban, Nguyen Tien, & Han, Nguyen Dinh. "An advanced energy efficient and high performance routing protocol for MANET in 5G." *Journal of Communication* 13.12 (2018): 743–749.

[40] Sheetal, J. "Architecture of 5G technology in mobile communication." In *Proceedings of 18th IRF International Conference*, DigitalXplore, 2015, 11th January.

[41] Qiao, Shuaiqing, et al. "Hybrid seismic-electrical data acquisition station based on cloud technology and green IoT." *IEEE Access* 8 (2020): 31026–31033.

[42] Li, Bin, Fei, Zesong, & Zhang, Yan. "UAV communications for 5G and beyond: Recent advances and future trends." *IEEE Internet of Things Journal* 6.2 (2018): 2241–2263.

[43] Wang, Johnson JH. "Stealth communication via smart ultra-wide-band signal in 5G, radar, electronic warfare, etc." In *2020 IEEE International Symposium on Antennas and Propagation and North American Radio Science Meeting*. IEEE, 2020.

[44] Sartipi, Farid. "Influence of 5G and IoT in construction and demolition waste recycling-conceptual smart city design." *Journal of Construction Materials* 1 (2020): 4–1.

[45] Yang, Dongmei, et al. "5G mobile communication convergence protocol architecture and key technologies in satellite internet of things system." *Alexandria Engineering Journal* 60.1 (2021): 465–476.

[46] D'Errico, Leonardo, et al. "Structural health monitoring and earthquake early warning on 5G uRLLC network." In *2019 IEEE 5th World Forum on Internet of Things (WF-IoT)*. IEEE, 2019.

[47] Khan, I. U., Abdollahi, A., Alturki, R., Alshehri, M. D., Ikram, M. A., Alyamani, H. J., & Khan, S. "Intelligent detection system enabled attack probability using markov chain in aerial networks." *Wireless Communications and Mobile Computing* 2021 (2021): 1–9.

[48] Johnson, David B., & Maltz, David A. "Dynamic source routing in ad hoc wireless networks." In *Mobile computing* (pp. 153–181). Springer, 1996.

[49] Perkins, Charles E., & Royer, Elizabeth M. "Ad-hoc on-demand distance vector routing." In *Proceedings WMCSA'99. Second IEEE Workshop on Mobile Computing Systems and Applications* (pp. 90–100). IEEE, 1999.

[50] Singh, Ankit, Singh, Gurpreet, & Singh, Mandeep. "Comparative study of OLSR, DSDV, AODV, DSR and ZRP routing protocols under blackhole attack in mobile ad hoc network." In *Intelligent Communication, Control and Devices* (pp. 443–453). Springer, Singapore, 2018.

[51] Yang, Hua, Li, Zhimei, & Liu, Zhiyong. "A method of routing optimization using CHNN in MANET." *Journal of Ambient Intelligence and Humanized Computing* 10.5 (2019): 1759–1768.

[52] Yang, Hua, Li, Zhimei, & Liu, Zhiyong. "Neural networks for MANET AODV: An optimization approach." *Cluster Computing* 20.4 (2017): 3369–3377.

[53] Khan, I. U., Hassan, M. A., Fayaz, M., Gwak, J., & Aziz, M. A. "Improved sequencing heuristic DSDV protocol using nomadic mobility model for FANETS." *CMC-Computers, Materials & Continua* 70.2 (2022): 3653–3666.

[54] Perkins, Charles E., & Bhagwat, Pravin. "Highly dynamic destination-sequenced distance-vector routing (DSDV) for mobile computers." *ACM SIGCOMM Computer Communication Review* 24.4 (1994): 234–244.

[55] T. Clausen, P. Jacquet, *Optimized link state routing protocol (OLSR) (No. RFC 3626)* (2003).

[56] Wu, Yi, Xu, Lei, Lin, Xiao, & Fang, Jie. "A new routing protocol based on OLSR designed for UANET maritime search and rescue." In *International Conference on Internet of Vehicles* (pp. 79–91). Springer, 2017.

[57] Haas, Z. J., Pearlman, M. R., & Samar, P. *The Zone Routing Protocol (ZRP) for Ad Hoc Networks* (2002). https://cir.nii.ac.jp/crid/1574231874964478720.

[58] AlKhatieb, A., Felemban, E., & Naseer, A. "Performance evaluation of ad-hoc routing protocols in (FANETs)." In *2020 IEEE Wireless Communications and Networking Conference Workshops (WCNCW)* (pp. 1–6). IEEE, 2020, April.

[59] Mariyappan, K., Christo, M. S., & Khilar, R. (2021). Implementation of FANET energy efficient AODV routing protocols for flying ad hoc networks [FEEAODV]. *Materials Today: Proceedings*, Elsevier.

[60] da Costa, L. A. L., Kunst, R., & de Freitas, E. P. "Q-FANET: Improved Q-learning based routing protocol for FANETs. *Computer Networks* 198 (2021): 108379.

[61] QasMarrogy, G. "Evaluation of flying ad hoc network topologies, mobility models, and IEEE standards for different video applications." *ARO-The Scientific Journal of Koya University* 9.1 (2021): 77–88.

[62] Bautista, O., Akkaya, K., & Uluagac, A. S. Customized novel routing metrics for wireless mesh-based swarm-of-drones applications. *Internet of Things* 11 (2020): 100265.

[63] Darabkh, K. A., Alfawares, M. G., & Althunibat, S. "MDRMA: Multi-data rate mobility-aware AODV-based protocol for flying ad-hoc networks. *Vehicular Communications* 18 (2019): 100163.

[64] Leonov, A. V. Application of bee colony algorithm for FANET routing. In *2016 17th International Conference of Young Specialists on Micro/Nanotechnologies and Electron Devices (EDM)* (pp. 124–132). IEEE, 2016, June.

[65] Gupta, N. K., Yadav, R. S., & Nagaria, R. K. "3D geographical routing protocols in wireless ad hoc and sensor networks: An overview." *Wireless Networks* 26.4 (2020): 2549–2566.

[66] del Peral-Rosado, J. A., Raulefs, R., López-Salcedo, J. A., & Seco-Granados, G. Survey of cellular mobile radio localization methods: From 1G to 5G. *IEEE Communications Surveys & Tutorials* 20.2 (2017): 1124–1148.

[67] Ibrahim, M. M. S., & Shanmugaraja, P. "Optimized link state routing protocol performance in flying ad-hoc networks for various data rates of unmanned aerial network." *Materials Today: Proceedings* 37 (2021): 3561–3568.

[68] Taher, H. M. "Sea Ad hoc Network (SANET) challenges and development." *International Journal of Applied Engineering Research* 13.8 (2018): 6257–6261.

[69] Osseiran, Afif, Monserrat, Jose F., & Marsch, Patrick eds. *5G mobile and wireless communications technology.* Cambridge University Press, 2016.

[70] Vora, L. J. Evolution of mobile generation technology: 1G to 5G and review of upcoming wireless technology 5G. *International Journal of Modern Trends in Engineering and Research* 2.10 (2015): 281–290.

[71] Jang, K., Han, M., Cho, S., Ryu, H. K., Lee, J., Lee, Y., & Moon, S. B. 3G and 3.5G wireless network performance measured from moving cars and high-speed trains. In *Proceedings of the 1st ACM Workshop on Mobile Internet through Cellular Networks* (pp. 19–24), ACM, 2009, September.

[72] Gupta, A., & Jha, R. K. "A survey of 5G network: Architecture and emerging technologies." *IEEE Access* 3 (2015): 1206–1232.

[73] Akyildiz, I. F., Gutierrez-Estevez, D. M., & Reyes, E. C.. The evolution to 4G cellular systems: LTE-Advanced. *Physical Communication* 3.4 (2010): 217–244.

[74] Shafi, M., Molisch, A. F., Smith, P. J., Haustein, T., Zhu, P., De Silva, P., ... & Wunder, G. 5G: A tutorial overview of standards, trials, challenges, deployment, and practice. *IEEE Journal on Selected Areas in Communications* 35.6 (2017): 1201–1221.

[75] Qadir, Q. M., Rashid, T. A., Al-Salihi, N. K., Ismael, B., Kist, A. A. & Zhang, Z. "Low power wide area networks: A survey of enabling technologies, applications and interoperability needs." *IEEE Access* 6 (2018): 77454–77473. https://doi.org/10.1109/ACCESS.2018.2883151.

[76] Abdollahi, A., & Fathi, M. "An intrusion detection system on ping of death attacks in IoT networks." *Wireless Personal Communication* 112 (2020): 2057–2070. https://doi.org/10.1007/s11277-020-07139-y.

[77] Raza, U., Parag, K., and Mahesh, S. "Low power wide area networks: An overview." *IEEE Communications Surveys & Tutorials* 19.2 (2017): 855–873.

Chapter 6

LPWAN technology for demand side management

Zupash
COMSATS University Islamabad

Abdul Qahar Shahzad
Quaid-i-Azam University

Farhood Nishat
Institute of System Engineering, Riphah University

Muhammad Allah Rakha
FAST National University of Computer and Emerging Sciences

Muhammad Yaseen Ayub
COMSATS University Islamabad

6.1 INTRODUCTION

Energy fields are facing a few great hurdles, which include an increasing demand for energy, previous infrastructure, and environmental concerns. To manage those problems, the idea of call for aspect control (DSM) has emerged as a promising answer for optimizing strength consumption in exceptional settings, including homes, corporations, and factories. DSM involves using various strategies and technologies to encourage people to use electricity during those specific hours when the demand and cost are lower [1]. One of the most popular technology for enabling DSM is the Low Power Wide Area Network (LPWAN) technology. This wireless communication system can transmit small amounts of data over long distant areas by consuming low power [2]. DSM is crucial for the electricity grid since it enables the management of energy demand in response to supply and pricing constraints. LPWAN technology has shown significant capacity for DSM, since it allows long-range communication with low power consumption, thus making it perfect for devices that need to transmit data over long distances at the same time as ultimate battery powered. LPWAN era can also help in applying diverse sensors and smart meters that provide real-time records on strength consumption, permitting utilities to

DOI: 10.1201/9781003426974-6

make more knowledgeable selections on electricity supply [3]. LPWAN technology for DSM has numerous benefits including improved energy efficiency, lower energy costs, and an effective demand response management. Real-time monitoring of energy consumption patterns with LPWAN technology can detect energy usage inefficiencies by allowing utilities to develop the focused energy management strategies. Additionally, LPWAN technology can also help in the implementation of dynamic pricing schemes that encourage people to reduce the consumption of energy which results in reduced energy costs for consumers and a more efficient use of the electricity grid [4]. This book chapter explores the use of LPWAN technology for DSM. It covers an overview of LPWAN technology and its benefits for DSM. It also discusses different types of LPWAN technology, their architecture and communication protocols for DSM. Furthermore, it also gives information about the applications of LPWAN technology for DSM in different settings such as homes, businesses, factories, and its integration with smart grid systems.

6.2 LITERATURE REVIEW

Low-Power WAN (LPWANs) have emerged as a promising solution for energy management across various sectors, including the energy industry. These networks provide a reliable and cost-effective communication system that enables the integration of sensors and devices for efficient energy management. The first section of this chapter is an introduction to LPWAN technology for demand of side management, emphasizing its benefits and significance in managing the consumption of energy. The second part of this chapter covers different types of LPWAN technology including Lora WAN, Sigfox, and NB-IoT with a detailed discussion on their architecture, communication protocols and devices/sensors used in the demand side management (DSM). The third section explains the concept of DSM, challenges and opportunities associated with it, and tells about the demand response programs and load shifting techniques used in the energy sector. The fourth section of this chapter focuses on the practical applications of LPWAN technology across the residential, commercial sectors as well as its integration with smart grid systems for the optimization. The fifth section presents real-world examples of LPWAN technology with the use cases and success stories. Finally, end part discusses the emerging trends and technologies in LPWAN for DSM, future prospects along with the challenges. LPWAN technology provides numerous benefits, such as cost savings, increased efficiency and less environmental impact along with its potential for implementation.

6.3 LPWAN TECHNOLOGY FOR DSM

Low Power WAN (LPWAN) technology has emerged as a highly effective solution for managing energy demand. It provides long-range connectivity, consumes less power and is also cost-effective. All these abilities make it suitable for various applications. Lora WAN, Sigfox, and NB-IoT are several types of this technology [5].

6.3.1 Lora WAN

Lora WAN (Long Range Wide Area Network) is a type of LPWAN technology. Lora WAN uses chirp spread spectrum modulation which makes it ideal for applications that require long-range connectivity, including remote monitoring and smart cities. It is important to note that Lora and Lora WAN are not interchangeable terms as they operate on different layers of the network. LoRa works on the physical layer while Lora WAN operates on the MAC layer, as shown in Figure 6.1.

The Lora WAN network architecture follows a topology named as star topology. In this network, the end devices are connected to the gateways directly rather than being connected through other nodes. This reduces the complexity of the network and increases its capacity. The power efficiency of the nodes enhances the elimination of unnecessary propagation between the nodes. The four key components of the Lora WAN architecture are end nodes, gateways, network servers, and application servers [6].

6.3.2 Sigfox

Sigfox is a type of LPWAN technology that operates on the ISM (Industrial, Scientific, and Medical) frequency band. It uses unique modulation method to provide connectivity over long distances. Sigfox is suitable for applications that require low data rates and low power consumption for example smart agriculture and asset tracking. Its protocol stack consists of four layers: application, physical, MAC and frequency band layers. The frequency

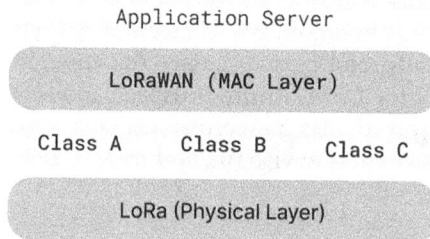

```
              Application Server

     ┌──────────────────────────────────┐
     │       LoRaWAN (MAC Layer)         │
     └──────────────────────────────────┘

       Class A      Class B      Class C

     ┌──────────────────────────────────┐
     │       LoRa (Physical Layer)       │
     └──────────────────────────────────┘
```

Figure 6.1 Lora WAN protocol stack.

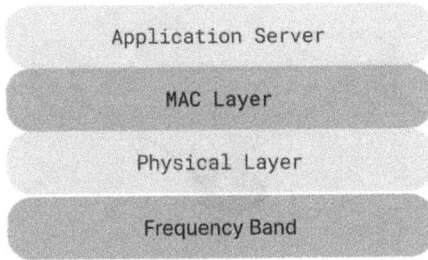

Figure 6.2 Sigfox protocol stack.

band layer manages the power needs and frequency allocation of the end nodes and gateways. The bodily layer is liable for embedding preambles sequence at the transmit stop and removing them at the receiver end. The MAC layer is in rate of managing MAC packets and constructing frames in line with the uplink and downlink formats. The software layer helps diverse programs. This generation is beneficial for packages that require low-velocity statistics transmission. Figure 6.2 shows the protocol stack of Sigfox [6,7].

6.3.3 NB-IoT

NB-IoT stands for Narrowband Internet of Things. It is a 3GPP standard for LPWAN technology. It operates on existing cellular networks which provides connectivity over long distances with low power consumption and low data rates. NB-IoT is suitable for applications that require secure and reliable connectivity. It uses a bandwidth of only 180 kHz, which results in lower data rates of about 250 kbps down-link and 20 kbps up-link. Though it is not well-suited for Fot A (Firmware-over-the-air) due to low data rates, it is more energy-efficient than other cellular technologies like eMTC and has a greater range. NB-IoT can be used in three different modes which are in-band, guard-band LTE, and standalone. It is already to be had in some regions and continues to be deployed in other nations. However, it isn't always appropriate for mobile IoT packages as it does not guide handover [6,7]. Figure 6.3 shows the architecture of NB-IoT [8].

6.3.4 LPWAN architecture for DSM

LPWAN technology architecture for DSM includes sensors, gateways, and cloud-based platforms. The gateways are responsible for receiving and transmitting data from sensors to the cloud-based platform. The sensors collect data from various sources such as from temperature sensors, humidity sensors, and smart meters, and transmit it to the gateway. Cloud-based platform collects and analyzes the data in order to provide insights into

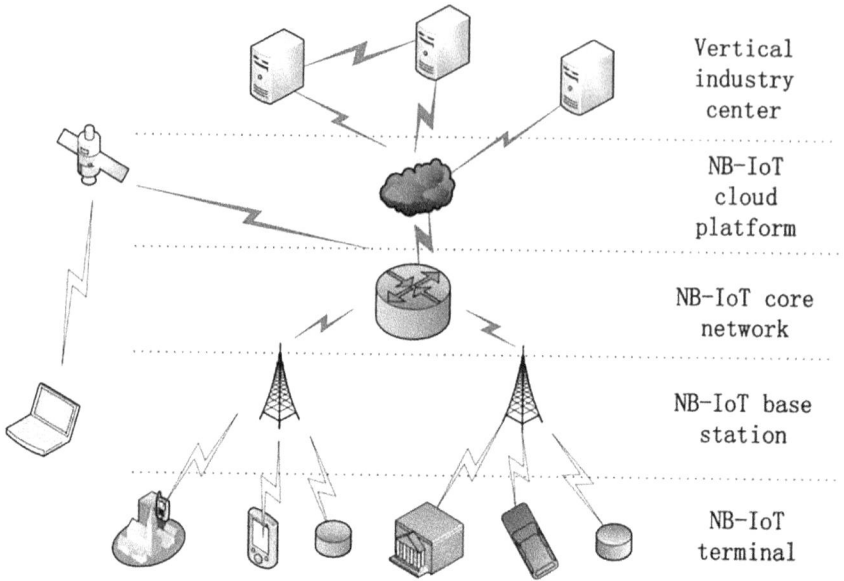

Figure 6.3 NB-IoT architecture.

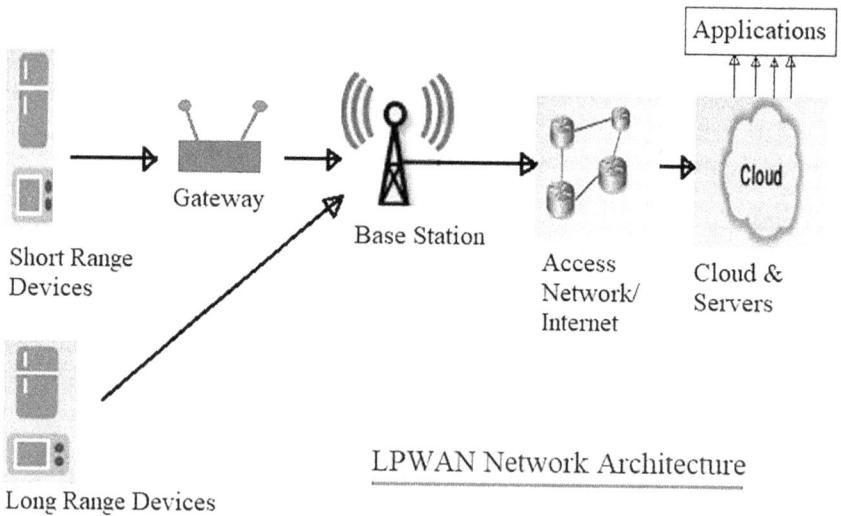

Figure 6.4 LPWAN network architecture.

energy consumption patterns and demand response programs. Figure 6.4 shows LPWAN technology architecture [9].

6.3.5 LPWAN-enabled devices and sensors for DSM

LPWAN-enabled gadgets and sensors are essential factors of DSM. These devices and sensors collect information and transmit it based on power consumption patterns, which might be used to pick out power-saving possibilities and call for response programs. LPWAN-based communication protocols, which include MQTT (Message Queuing Telemetry Transport) and CAP (Constrained Application Protocol) offers comfy and dependable connectivity among LPWAN-enabled devices and the cloud-based totally platform [6]. LPWAN technology offers many benefits for DSM including cost-effectiveness, low power consumption, and long-range connectivity. The use of LPWAN era in demand side control is anticipated to boom inside the destiny. This may be pushed by way of the emergence of the latest LPWAN technology and through the developing call for power-efficient solutions.

6.4 DSM IN ENERGY SECTOR

DSM is a hard and fast of technology, applications, and moves which are carried out on the demand side of electricity meters. These are generally applied with the aid of governments, utilities, third parties, or people. The reason of DSM is to manipulate and decrease energy intake through electricity efficiency, power conservation, demand response and on-site technology and storage. DSM ambitions to reduce the overall strength of machine expenses and accomplish coverage goals facet by means of facet, which includes reducing greenhouse fuel emissions, balancing deliver and demand, or reducing consumer electricity payments, implement various technologies and strategies that can either shift the timing or reduce the level of energy consumption in response to changing energy prices or system conditions, as a result improving the efficiency of the power system and reducing the need for expensive peak capacity [10].

Figure 6.5 illustrates that DSM encompasses several categories such as incentive payment-based demand response, on-site generation, energy efficiency, energy conservation, price-based demand response, and on-site storage.

6.4.1 Challenges and opportunities in DSM

DSM is regarded as a potential technique to increase energy efficiency and to decrease greenhouse gas emissions in energy sector. However, DSM also faces various difficulties that need to be fixed to make the most its potential. One of the major challenges faced by DSM is the lack of awareness and knowledge about DSM among people and businesses which can make it difficult to encourage their participation in DSM programs and initiatives.

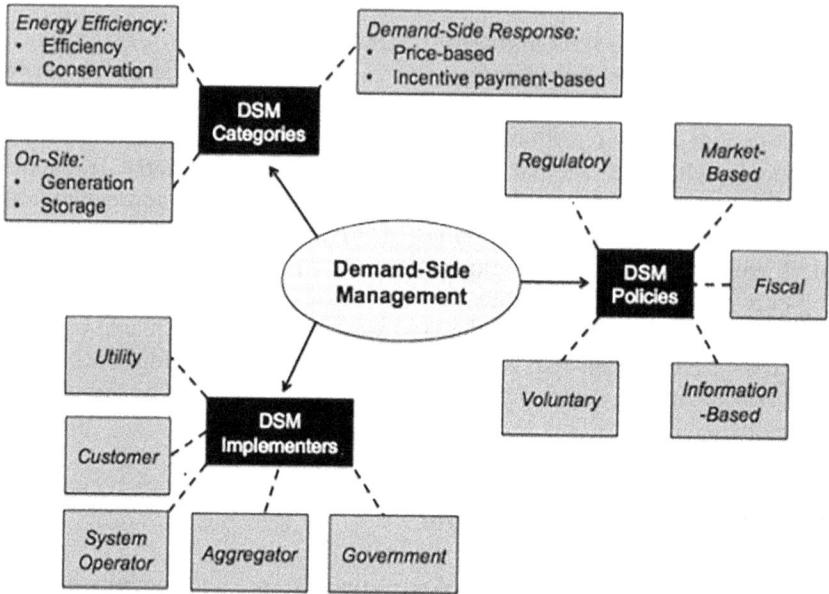

Figure 6.5 Categorizing DSM into three parts—policies, implementers, and categories.

Another challenge is the absence of policy and regulatory frameworks that aid the implementation of DSM especially in developing countries [11]. However, DSM offers several opportunities, which include reducing energy costs, enhancing grid reliability, and promoting sustainability. DSM also helps to tackle the issue of peak demand which can strain the power grid and cause blackouts or brownouts. DSM policies supports in order to achieve policy objectives such as reducing greenhouse gas emissions and balancing supply and demand. Furthermore, the implementation of DSM can create new business opportunities and markets which also helps in generating employment [12].

6.4.2 Demand response programs and load shifting techniques

DR packages and cargo shifting strategies are the two vital ways for coping with power needs. DR applications help to alter electricity usage according to adjustments in fees and grid situations. Load shifting strategies contain the use of strength in the course of off-top durations for you to reduce the burden during peak hours which enables to enhance the reliability of the power grid [13]. DR programs and load shifting techniques are important ways to manage energy demand and improve energy efficiency. DR programs involve changing the way electricity is used in response to

Load shape	DR type
	Load Shifting
	Flexible load shapes (dynamic energy management)
	Strategic Conservation (energy efficiency)

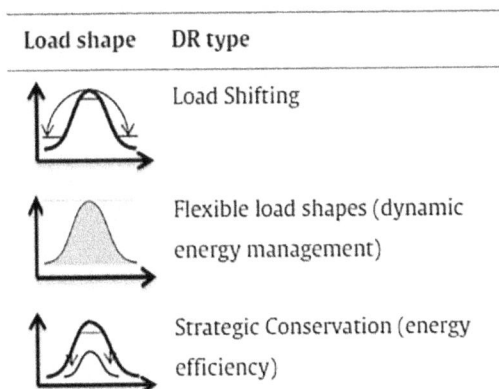

Figure 6.6 Load shapes resulting from DSM.

do changings in pricing and grid conditions. This includes direct control of electricity usage by utilities, paying higher prices during peak periods or different pricing at different times of day to encourage off-peak usage. Load shifting techniques involves adjusting energy consumption patterns to reduce peak demand such as using energy-efficient appliances or adjusting thermostats. These strategies can help improve grid reliability and reduce greenhouse gas emissions [14,15].

Figure 6.6 shows that demand response (DR) can be applied in different ways to adjust the traditional electricity demand. It involves the modification of consumption patterns in response to changes in electricity prices and grid conditions, shifting electricity demand from peak periods to off-peak periods, and allowing utilities to directly control the electricity usage of people during peak demand periods. DR can be used as a strategy to optimize energy usage to improve grid reliability and to reduce greenhouse gas emissions [16].

6.5 APPLICATIONS OF LPWAN TECHNOLOGY FOR DSM

LPWAN era is a crucial device for powerful DSM. It may be used in residential settings in an effort to manage and reveal power usage through smart thermostats and equipment management systems. This allows people to alter their power utilization patterns to reduce waste and fees. Commercial and business settings can also benefit from LPWAN generation in a way for constructing automation and power management, leading to good sized strength financial savings. This technology can also be integrated with smart grid systems to optimize energy usage and DR. Utilities can analyze energy usage patterns and develop more effective DR programs while efficiently managing energy supply by collecting real-time data [17–20].

6.5.1 Residential DSM applications

Residential call for side management (DSM) technologies and strategies are applied to control and lower energy consumption in homes. Smart thermostats are an instance of those technology which enable owners to remotely manage their heating and cooling structures and set schedules for energy conservation. Lighting controls and equipment control systems also can offer real-time power intake records to house owners helping them lessen strength waste and costs. However, the main purpose of residential DSM applications is to assist homeowners in saving energy and money while also contributing to reducing greenhouse gas emissions [17].

6.5.2 Commercial and industrial DSM applications

Commercial and business demand facet control (DSM) technologies and techniques are used to manipulate and decrease electricity consumption in organizations, factories, and other commercial settings, such as constructing automation systems, and it could control heating, cooling, and lighting fixtures systems to enhance strength efficiency. Moreover, strength management systems can screen and optimize energy consumption in actual time and other DSM applications include electricity-green lights, variable frequency drives, and power component correction structures. These technologies can assist corporations and industries to shop on strength costs and additionally beautify their average strength efficiency [18].

6.5.3 Smart grid integration and optimization

Smart grid integration and optimization is a vital issue of demand aspect management (DSM). It facilitates utilities and people to manipulate electricity delivery and its call for more efficiently. LPWAN generation can be included with smart grid systems to acquire real-time information and examine strength utilization patterns. These statistics can tell greater powerful demand reaction programs and help utilities in adjusting electricity supply to meet changing demand. This method reduces the danger of blackouts and improves the general reliability of the strength grid. Smart grid integration can also enable other DSM applications, inclusive of allotted energy sources (DERs). These are small-scale strength technology gadgets that can be included into the power grid to offer extra strength all through height demand durations. LPWAN technology can display and manage these DERs in real time, while optimizing their use and ensuring that they're handiest used when wanted. In popular use, clever grid integration and optimization are crucial for effective call for facet control. It reduces strength waste, lowers costs, and helps utilities and consumers paintings collectively to manage power deliver and demand [19,20].

6.6 CASE STUDIES AND EXAMPLES

LPWAN technology is being used in many real-world applications in order to optimize DSM and also helps in reducing energy waste and costs in various industries. One of the most prominent examples is the implementation of smart street lighting systems in the city of Barcelona. This system uses LPWAN technology to monitor and control energy usage in real-time. The smart lighting system adjusts the lighting intensity based on its surrounding light conditions and pedestrian traffic patterns, which reduces energy usage and has resulted in significant cost reductions for the city [21]. In the agriculture industry, LPWAN technology has been used to monitor and optimize irrigation systems. Farmers can use this technology to remotely control irrigation systems based on real-time data on weather and soil moisture, which reduces water waste and improves crop yields [22]. This technology is also being used in the healthcare industry to remotely monitor patient health, which improves the efficiency of medical equipment usage. Such medical devices that use LPWAN technology can transmit real-time data on patients, vital signs to medical professionals, allowing for more timely interventions and improved patient outcomes. In general, LPWAN technology has shown great potential for demand in various industries with many success stories and use cases demonstrating its effectiveness in reducing energy waste and costs while improving overall efficiency [23,24].

6.7 SECURITY AND THREAT MANAGEMENT IN LPWAN FOR DSM

In the domain of LPWAN technology for DSM, ensuring the security and integrity of the network infrastructure is very important. As the implementation of LPWAN networks expands to include aerial networks, such as drones and unmanned aerial vehicles (UAVs), the need for strong security measures becomes even more difficult. One specific area of concern is the detection and mitigation of potential attacks within these aerial networks. To address this challenge, an intelligent detection system enabled by Markov Chain analysis is developed as a promising strategy to assess the attack probability and enhance the security of LPWAN in aerial networks [25].

As the implementation of LPWAN networks continues to increase, the possible weaknesses and threats associated with IoT networks become increasingly clear. Specific issue is "Ping of Death Attacks," which can disturb the operation of IoT devices connected to LPWAN networks. To minimize the risk and strengthen the security measures, the implementation of an advanced Intrusion Detection System (IDS) has emerged as a vital element. An IDS specifically designed for LPWAN networks enables the early recognition and response to potential Ping of Death Attacks, safeguarding the integrity and availability of the connected devices. By utilizing the strength of advanced algorithms and real-time monitoring, the IDS can detect and alert network administrators to suspicious activities representative of Ping of Death Attacks, therefore allowing for timely preventive measures and reducing the potential impact [26].

6.8 FUTURE OF LPWAN TECHNOLOGY FOR DSM

LPWAN era has proved itself as a precious device in DSM and is expected to be more crucial within the destiny of strength control. There are numerous emerging traits and technology in LPWAN for DSM which can be worth examining. One of these rising developments is the mixing of LPWAN generation with synthetic intelligence (AI) and gadget learning (ML) algorithms. These algorithms analyze large0020amounts of statistics collected through LPWAN gadgets to pick out patterns and expect energy consumption trends. This will allow utilities and people to better manipulate strength deliver and call for, to optimize strength utilization, and to lessen strength waste. Another rising technology is the use of LPWAN gadgets for asset management and predictive maintenance. Maintenance needs can be recognized and addressed earlier than equipment failures occur by remotely monitoring equipment and reading information amassed through LPWAN devices. This technique will lessen downtime, will lower maintenance costs, and additionally improve usual gadget reliability [27,28].

Regardless of these encouraging developments, there are also potential future developments and challenges that may affect the use of LPWAN technology for DSM. Such as the adoption of renewable energy sources like solar and wind power may increase. LPWAN technology can help to manage the variability of these energy sources by allowing real-time monitoring and control of energy supply and demand. On the other hand, there are cybersecurity risks and a need for interoperability between different LPWAN technologies that must be addressed as LPWAN technology continues to evolve and become more widespread. LPWAN technology can continue to improve energy efficiency, reduce energy waste and optimize energy usage by making it a crucial tool for DSM in the future by overcoming these challenges [29–32].

6.9 CONCLUSION

LPWAN technology has ended up being a treasured device for handling power intake through demand facet control. Its low electricity utilization, long-variety connectivity, and cost-effectiveness had made it a perfect answer for actual time tracking and management of strength consumption. It is likewise key to the advanced reliability and efficiency of energy systems. This chapter included numerous styles of LPWAN era, their architecture, conversation protocols, and devices used in call for aspect control in addition to the importance of demand aspect control within the power region. Effective demand reaction packages and load shifting techniques were also discussed. Moreover, the bankruptcy had highlighted a hit LPWAN implementation in residential, commercial, and business sectors, as well as smart grid integration and optimization. It is obvious that LPWAN era will play an increasingly vital role in the future of call for aspect management, and it'll preserve to conform and will meet the strength region's ever-developing demands.

REFERENCES

[1] Enright, T. O. N. I. & Faruqui, A. (2012). A bibliography on dynamic pricing of electricity. *The Battle Group*, Cambridge.
[2] Adefemi Alimi, K. O., Ouahada, K., Abu-Mahfouz, A. M., & Rimer, S. (2020). A survey on the security of low power wide area networks: Threats, challenges, and potential solutions. *Sensors, 20*(20), 5800.
[3] Bakare, M. S., Abdulkarim, A., Zeeshan, M., & Shuaibu, A. N. (2023). A comprehensive overview on demand side energy management towards smart grids: Challenges, solutions, and future direction. *Energy Informatics, 6*(1), 1–59.
[4] Deng, R., Yang, Z., Chow, M. Y., & Chen, J. (2015). A survey on demand response in smart grids: Mathematical models and approaches. *IEEE Transactions on Industrial Informatics, 11*(3), 570–582.
[5] Albeyboni, H. J. & Ali, I. A. (2023). LPWAN technologies for IOT applications: A review. *Journal of Duhok University, 26*(1), 29–42. Retrieved from https://journal.uod.ac/index.php/uodjournal/article/view/2298.
[6] Foubert, B. & Mitton, N. (2020). Long-range wireless radio technologies: A survey. *Future Internet, 12*(1), 13.
[7] Al-Sarawi, S., Anbar, M., Alieyan, K., & Alzubaidi, M. (2017, May). Internet of Things (IoT) communication protocols. In *2017 8th International conference on information technology (ICIT)* (pp. 685–690). IEEE.
[8] Nashiruddin, M. I., & Purnama, A. A. F. (2020, June). Nb-iot network planning for advanced metering infrastructure in Surabaya, Sidoarjo, and Gresik. In *2020 8th International Conference on Information and Communication Technology (ICoICT)* (pp. 1–6). IEEE.

[9] Song, Y., Lin, J., Tang, M., & Dong, S., (2017). An Internet of energy things based on wireless LPWAN. *Engineering*, *3*(4), 460–466.

[10] Warren, P., (2015). Demand-side management policy: Mechanisms for success and failure (Doctoral dissertation), University College London.

[11] International Energy Agency, (2014). Capturing the multiple benefits of energy efficiency. OECD.

[12] Rezaei, N., Ahmadi, A., & Deihimi, M., (2022). A comprehensive review of demand-side management based on analysis of productivity: Techniques and applications. *Energies*, *15*(20), 7614.

[13] Wang, Z., Paranjape, R., Sadanand, A., & Chen, Z., (2013, May). Residential demand response: An overview of recent simulation and modeling applications. In *2013 26th IEEE Canadian Conference on Electrical and Computer Engineering (CCECE)* (pp. 1–6). IEEE.

[14] Dharani, R., Balasubramonian, M., Babu, T.S., & Nastasi, B., (2021). Load shifting and peak clipping for reducing energy consumption in an Indian university campus. *Energies*, *14*(3), 558.

[15] Lyden, A., Pepper, R., & Tuohy, P.G., (2018). A modelling tool selection process for planning of community scale energy systems including storage and demand side management. *Sustainable Cities and Society*, *39*, 674–688.

[16] Chuang, A.S. & Gellings, C.W., (2008). Demand-side integration in a restructured electric power industry. CIGRE, number Paper C6-105, Session.

[17] Iqbal, S., Sarfraz, M., Ayyub, M., Tariq, M., Chakrabortty, R.K., Ryan, M.J., & Alamri, B., (2021). A comprehensive review on residential demand side management strategies in smart grid environment. *Sustainability*, *13*(13), 7170.

[18] Gungor, V.C. & Hancke, G.P., (2009). Industrial wireless sensor networks: Challenges, design principles, and technical approaches. *IEEE Transactions on Industrial Electronics*, *56*(10), 4258–4265.

[19] Sioshansi, F. ed., (2011). *Smart grid: Integrating renewable, distributed and efficient energy.* Academic Press.

[20] Kundur, P.S. & Malik, O.P., (2022). *Power system stability and control.* McGraw-Hill Education.

[21] Chilamkurthy, N.S., Pandey, O.J., Ghosh, A., Cenkeramaddi, L.R. & Dai, H.N., (2022). *Low-power wide-area networks: A broad overview of its different aspects.* IEEE Access.

[22] Balamurugan, S., Divyabharathi, N., Jayashruthi, K., Bowiya, M., Shermy, R.P. & Shanker, R., (2016). Internet of agriculture: Applying IoT to improve food and farming technology. *International Research Journal of Engineering and Technology (IRJET)*, *3*(10), 713–719.

[23] Adi, P.D.P. & Kitagawa, A., (2020, August). Performance evaluation of low power wide area (LPWA) LoRa 920 MHz sensor node to medical monitoring IoT based. In *2020 10th Electrical Power, Electronics, Communications, Controls and Informatics Seminar (EECCIS)* (pp. 278–283). IEEE.

[24] Catherwood, P.A., Steele, D., Little, M., Mccomb, S. & McLaughlin, J., (2018). A community-based IoT personalized wireless healthcare solution trial. *IEEE Journal of Translational Engineering in Health and Medicine*, *6*, 1–13.

[25] Khan, I.U., Abdollahi, A., Alturki, R., Alshehri, M.D., Ikram, M.A., Alyamani, H.J. & Khan, S., (2021). Intelligent detection system enabled attack probability using Markov chain in aerial networks. *Wireless Communications and Mobile Computing*, 2021, 1–9.

[26] Abdollahi, A. & Fathi, M., 2020. An intrusion detection system on ping of death attacks in IoT networks. *Wireless Personal Communications*, 112, 2057–2070.

[27] Matni, N., Moraes, J., Pacheco, L., Rosário, D., Oliveira, H., Cerqueira, E. & Neto, A., (2020, June). Experimenting long range wide area network in an e-health environment: Discussion and future directions. In *2020 International Wireless Communications and Mobile Computing (IWCMC)* (pp. 758–763). IEEE.

[28] Xiong, X., Zheng, K., Xu, R., Xiang, W. & Chatzimisios, P., (2015). Low power wide area machine-to-machine networks: Key techniques and prototype. *IEEE Communications Magazine*, 53(9), 64–71.

[29] Mekki, K., Bajic, E., Chaxel, F. & Meyer, F., (2019). A comparative study of LPWAN technologies for large-scale IoT deployment. *ICT Express*, 5(1), 1–7.

[30] Al-Fuqaha, A., Guizani, M., Mohammadi, M., Aledhari, M. & Ayyash, M., (2015). Internet of things: A survey on enabling technologies, protocols, and applications. *IEEE Communications Surveys & Tutorials*, 17(4), 2347–2376.

[31] Kumar, S., Tiwari, P. & Zymbler, M., (2019). Internet of Things is a revolutionary approach for future technology enhancement: A review. *Journal of Big Data*, 6(1), 1–21.

[32] Mariano-Hernández, D., Hernández-Callejo, L., Zorita-Lamadrid, A., Duque-Pérez, O. & García, F.S., (2021). A review of strategies for building energy management system: Model predictive control, demand side management, optimization, and fault detect & diagnosis. *Journal of Building Engineering*, 33, 101692.

Chapter 7

AI-enabled low-powered wireless area networks for quality air

Ubaid Ullah and Muhammad Usama
University of Wah

Zaid Muhammad
Abdul Wali Khan University

Abdullah Akbar
National University of Computer and Emerging Sciences

7.1 INTRODUCTION

The process of connecting a variety of separate computer systems and electronic gadgets to one another over the internet is referred to as the "Internet of Things" (IoT), which is also an acronym [1]. This makes it possible for gadgets to converse and exchange data. One of the most useful uses of the IOT has been monitoring the environment, particularly the quality of the air [2]. Given rising concerns about air pollution and its harmful impacts on human health, it is crucial to monitor the air quality in diverse urban and industrial areas [3]. IoT-based environmental monitoring systems utilize it frequently [4]. LPWAN networks are useful for environmental monitoring applications because they are designed to transmit data over very long distances while consuming very little power [5].

AI-enabled low-powered wireless area networks (LPWANs) are an emerging technology that combines the benefits of low-power wireless communication and artificial intelligence (AI) to enable efficient and intelligent connectivity for a wide range of applications. LPWANs are designed to provide long-range, low-power communication between devices, making them suitable for IoT deployments where battery life and connectivity range are critical. Incorporating AI into LPWANs enhances their capabilities by enabling intelligent decision-making, data analysis, and optimization. Here are some critical aspects of AI-enabled LPWANs. AI algorithms can optimize the routing of data packets within the LPWAN, considering factors such as signal strength, network congestion, and device location. This ensures efficient data transmission and minimizes energy consumption [6]. AI can dynamically adjust the transmission power of LPWAN devices based

DOI: 10.1201/9781003426974-7

on real-time environmental conditions. AI algorithms can optimize power consumption without compromising network performance by analyzing signal quality, interference, and distance [7]. AI algorithms can analyze sensor data collected from LPWAN-connected devices to detect patterns and anomalies. By monitoring device health and performance, AI-enabled LPWANs can predict maintenance requirements and alert operators before failures occur, reducing downtime and maintenance costs [8]. AI can process the vast amount of data generated by LPWAN devices and extract meaningful insights. LPWANs can identify data patterns, trends, and correlations by applying machine learning algorithms, enabling predictive analytics and actionable intelligence [9,10].

An AI LPWAN, or artificial intelligence-enabled low-power wide area network, is a kind of wireless communication system designed to transfer modest amounts of data over great distances while using just a negligible amount of energy [11,12]. It is perfect for usage in IoT networks where several low-power devices must connect over an extensive area [13]. Low power consumption is one of the critical benefits of AI-enabled LPWAN technology [14]. This means that AI-enabled LPWAN devices can operate for long periods without needing frequent battery replacements. It also operates at low frequencies [13], which allows it to penetrate walls and other obstacles, making it suitable for indoor and outdoor applications. It is also cost-effective compared to other wireless communication technologies [15]. It requires fewer base stations to cover the same area, reducing infrastructure costs.

AI-enabled LPWAN technology monitors air quality well [16,17]. Remote monitoring is excellent for its low power consumption, long-range connectivity, and scalability. It may become increasingly crucial to providing clean air as air quality remains a critical public health issue [18]. This technology is perfect for IoT networks that monitor extensive regions or remote places, including air quality monitoring devices, due to its low power consumption, long-range connectivity, and cost-effectiveness. It permits vast numbers of low-cost sensors to monitor environmental parameters across a wide area, providing valuable insights to manage pollution and air quality challenges [19,20].

7.2 LITERATURE

Zanella et al.'s [21] survey article comprehensively overviews various LPWAN technologies and their potential applications in IoT systems. The authors first introduce the concept of IoT and its various applications, including environmental monitoring. They then discuss the critical challenges associated with IoT-based environmental monitoring, such as power consumption, range, data rate, and scalability. This chapter provides valuable insights for researchers, practitioners, and policymakers interested in developing and deploying AI-enabled LPWAN-based IoT systems [21]. Mehmood et al.'s [22] assessment covers current accomplishments and problems in IoT-based intelligent cities. The authors explore smart city needs and how IoT technology enables smart city applications. IoT technology enables smart city applications, the authors say. They show how IoT sensors and actuators can gather data on city factors such as traffic flow, energy usage, and air quality [23]. Wi-Fi, cellular, and LPWAN technologies can link these devices. The paper then describes IoT-based smart city applications such as smart traffic, parking, and energy management. Each application's pros, cons, and successful deployments are discussed [22].

Ahsan et al. discuss the implementation of an IoT system based on LoRa technology for monitoring and controlling household appliances. The authors propose a system that uses LoRa nodes for data transmission, a LoRa gateway for data aggregation and forwarding, and a cloud server for data storage and analysis presents an exciting application of LoRa technology for home automation, which could have significant benefits in terms of energy savings, convenience, and improved quality of life for users. However, it is essential to note that implementing such a system requires careful consideration of security and privacy concerns, particularly given the sensitive nature of the data being collected and transmitted [24]. Table 7.1 clearly indicates the techniques used, description, and limitations related to LPWAN technology.

Table 7.1 Standard techniques used in LPWAN, along with a description, limitations

References	Technique used	Description	Limitations
[25]	LoRaWAN	The paper presents a LoRaWAN-based wireless air quality monitoring system with nodes for sensing temperature, humidity, and PM2.5. The technology can provide real-time air quality data for metropolitan areas	The research does not consider the system's energy usage and the impact of interference on data transmission
[26]	NB-IoT	The paper describes a method for monitoring air quality using NB-IoT technologies. The system has CO_2, temperature, humidity, and PM2.5 measurement sensors. A cloud server receives the data for processing and display	The research does not evaluate the system's performance outdoors, where interference and other variables can impact data transmission

(Continued)

Table 7.1 (Continued) Standard techniques used in LPWAN, along with a description, limitations

References	Technique used	Description	Limitations
[27]	LoRaWAN	The research suggests a LoRaWAN-based air quality monitoring system that may be implemented cheaply. Carbon dioxide (CO_2), heat, dampness, and particulate matter (PM2.5) are all measured by the system's sensors. The technology is built to save energy and can run on batteries for long periods	The study does not account for the system's scalability or potential limitations of LoRaWAN technology in terms of network coverage and capacity
[28]	Sigfox	The purpose of this research is to provide a system for monitoring air quality that makes use of Sigfox technology. There are sensors in the system that can measure CO_2, temperature, humidity, and PM2.5. The information is sent to a cloud server for analysis and visualization. It is possible to run the system for lengthy periods using just the power provided by the batteries because of its energy-efficient design	The study doesn't look at how well the system works outside, where there may be interference and other things that make it hard to send data
[29]	LoRaWAN	The paper presents a LoRaWAN-based wireless air quality monitoring system with nodes for sensing temperature, humidity, and PM2.5. The system is intended to be energy-efficient and capable of running for lengthy periods on batteries. The research assesses the system's performance in an outdoor setting and proves its ability to provide real-time air quality data for metropolitan regions	The research does not consider the possible limits of LoRaWAN technology in terms of network coverage and capacity, especially in densely populated regions

(Continued)

Table 7.1 (Continued) Standard techniques used in LPWAN, along with a description, limitations

References	Technique used	Description	Limitations
[30]	LoRaWAN, NB-IoT, Sigfox, Weightless, and LTE-M	This paper examines the applications of several LPWAN technologies, including LoRaWAN, NB-IoT, Sigfox, Weightless, and LTE-M, in smart city scenarios. LoRaWAN and NB-IoT were the most prevalent LPWAN technologies for smart city applications, according to the study, but there was no universal solution	The report thoroughly evaluates numerous LPWAN technologies but does not provide a deep analysis of any one technology
[31]	LoRaWAN and NB-IoT	The performance of LoRaWAN and NB-IoT for smart city applications is compared in this research. According to the research, LoRaWAN is better-suitable for applications requiring long-range communication and low data rates. At the same time, NB-IoT is better suited for applications requiring more effective data rates and dependability	The performance of just two low-power wide-area network (LPWAN) technologies was tested in this research for the particular application of smart cities
[32]	LoRaWAN and Sigfox	This study compares the performance of Sigfox and LoRaWAN for use in smart cities. The research found that Sigfox was better suited for applications needing high reliability and low power consumption. At the same time, LoRaWAN was better suited for those that required long-range communication and low data rates	The performance of just two low-power wide-area network (LPWAN) technologies was tested in this research for the particular application of smart cities

This chapter explores the application of wireless-based IoT technologies using Low-Power Wide Area Networks (AI-enabled LPWAN) for monitoring and maintaining quality air. The three LPWAN technologies examined in this chapter are LoRaWAN, Sigfox, and NB-IoT. These technologies offer long-range, low-power connectivity, making them ideal for IoT applications that require extensive coverage and energy efficiency. This chapter delves into the principles and characteristics of each technology, discussing their advantages and limitations. It also highlights their potential to enable air quality monitoring systems, including sensor deployment, data collection, and communication protocols. By leveraging wireless-based IoT using AI-enabled LPWAN, this chapter presents an innovative approach to effectively monitor and improve air quality, leading to a healthier and more sustainable environment [33].

7.3 CRITICAL ASPECTS RELATED TO AI-ENABLED LPWAN TECHNOLOGY

7.3.1 AI in LPWAN deployment and optimization

Several studies use AI techniques to determine the optimal placement and configuration of LPWAN gateways to maximize coverage and minimize interference. AI algorithms, such as reinforcement learning and genetic algorithms, optimize the network's performance, reduce energy consumption, and improve scalability [34].

7.3.2 AI-driven data analytics in LPWAN

Data analytics for proactive maintenance: To identify abnormalities, forecast breakdowns, and allow proactive equipment maintenance, AI algorithms examine data from IoT devices linked to LPWAN networks. AI techniques, including machine learning algorithms, are applied to classify and analyze data generated by IoT devices connected to LPWAN networks, enabling intelligent decision-making and automation [35].

7.3.3 Security and privacy in AI-enabled LPWAN

AI-based intrusion detection: Machine learning algorithms are utilized to detect and mitigate security threats and attacks in LPWAN networks, enhancing network security. Privacy-preserving techniques: Researchers are investigating AI-driven approaches to ensure the privacy of user data transmitted over LPWAN networks, including encryption, anonymization, and differential privacy [36].

7.3.4 AI and energy efficiency in LPWAN

AI techniques are employed to design and optimize communication protocols that minimize power consumption while maintaining reliable connectivity in LPWAN networks. Algorithms are used to optimize energy harvesting techniques, such as solar or kinetic energy, and manage energy resources in IoT devices connected via LPWAN [37].

7.3.5 AI-enabled resource management in LPWAN

AI algorithms provide effective network resource allocation in LPWAN networks to serve a variety of IoT applications with varying needs. AI techniques can manage network congestion, optimize traffic flow, and prioritize critical data transmission in LPWAN networks [34].

7.4 APPLICATION AREAS OF AI-ENABLED LPWAN TECHNOLOGIES

AI-enabled LPWAN technologies have gained significant popularity in various industries because they provide long-range, low-power connectivity for IoT devices [38]. Here are some application areas where LPWAN technologies are commonly used:

7.4.1 Smart cities

AI LPWAN technologies are extensively used in innovative city applications for monitoring and controlling various infrastructure components. This includes intelligent street lighting, waste management systems, parking sensors, environmental monitoring, and smart metering for utilities such as water and electricity [39,40].

7.4.2 Industrial IoT

AI LPWAN technologies find applications in the manufacturing, logistics, and agriculture industries. They are used for asset tracking, supply chain management, predictive machinery maintenance, remote equipment monitoring, and optimizing operational efficiency [41,42].

7.4.3 Agriculture

AI LPWAN enables farmers to monitor and manage their crops and livestock more effectively. It is used for soil moisture monitoring, weather monitoring, precision irrigation, livestock tracking, and greenhouse environmental conditions [43].

7.4.4 Asset tracking

AI LPWAN provides a cost-effective solution for tracking and monitoring valuable assets such as vehicles, containers, and equipment. It allows businesses to track assets location, movement, and condition in real-time, enabling better inventory management and theft prevention [44,45].

7.4.5 Environmental monitoring

AI Low-Power Wide-Area Network technologies are used for the gathering and monitoring of environmental data, such as air quality, water quality, noise levels, and meteorological data [46]. This information is necessary for early warning systems, urban planning, and research [47].

7.4.6 Healthcare

AI LPWAN technologies have applications in healthcare for remote patient monitoring, tracking medical equipment, and managing inventory in hospitals and clinics. They allow healthcare professionals to offer individualized treatment, enhance patient outcomes, and effectively allocate resources [48,49].

7.4.7 Building automation

AI LPWAN technologies play a role in building automation systems for energy management, security, and occupancy monitoring. They enable efficient control and monitoring of lighting, HVAC systems, access control, and surveillance [50].

7.4.8 Retail

AI LPWAN is used in retail environments for inventory management, asset tracking, and customer analytics. It enables retailers to track and manage stock levels, monitor footfall and customer behavior, and optimize store layouts and displays [51].

7.4.9 Utilities

AI LPWAN technologies are utilized in utility sectors such as water, gas, and electricity for remote metering, consumption monitoring, and leak detection. They enable utilities to improve operational efficiency, detect anomalies, and reduce costs [52].

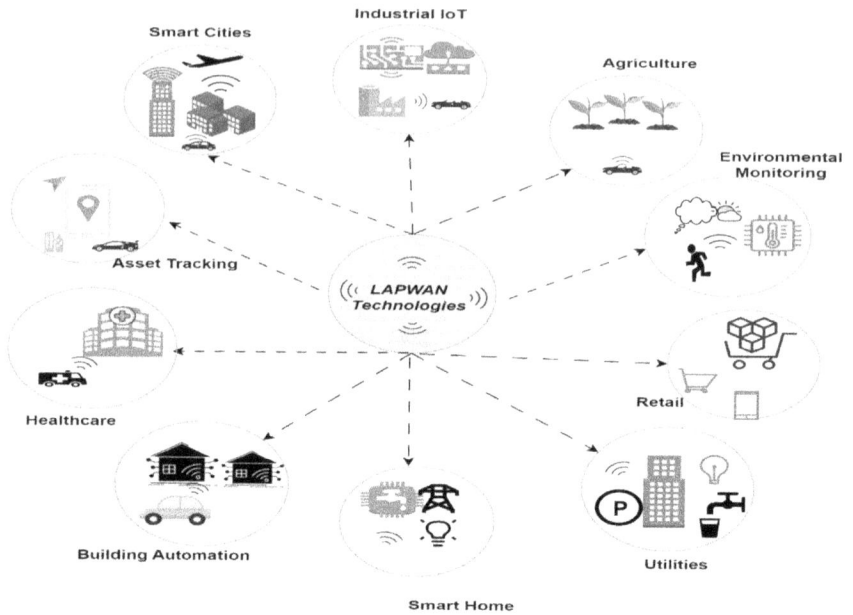

Figure 7.1 Application areas of AI-enabled LPWAN technologies.

7.4.10 Smart home

AI LPWAN technologies form the backbone of intelligent home systems, enabling the connectivity and control of various devices and appliances [44]. They facilitate home automation, energy management, security systems, and remote monitoring and control of home devices [53].

Figure 7.1 visually illustrates the diverse range of fields and sectors where AI-enabled LPWAN technologies find application, highlighting their versatility and potential impact on various industries.

7.5 IMPORTANCE OF MONITORING AIR QUALITY AND THE ROLE OF IoT NETWORKS

It is of the utmost importance to keep an eye on the air quality since poor air quality may have serious repercussions not just for human health but also for the environment [40]. Long-term exposure to high levels of air pollution has been linked not only to cancer and other problems that pose a risk to one's life but also to a variety of respiratory and cardiovascular diseases that have been demonstrated to be associated with the condition [54]. Damage to ecosystems, decreased agricultural yields, and climate change are just some of the detrimental effects that poor air quality has on the surrounding environment [55].

The provision of real-time data on pollutants and other environmental parameters via IoT networks may be of critical importance in the process of monitoring air quality [56]. These networks are made up of linked sensors that gather information on the air's temperature and humidity as well as the levels of a variety of pollutants, such as CO, NO_2, and PM [57]. Researchers and policymakers are able to obtain insights into the causes and effects of air pollution as well as devise effective methods for decreasing it when they analyze the data presented here. In addition, IoT networks make it possible to construct cutting-edge monitoring devices for air quality that are portable and can be set up in a variety of locations. For instance, mobile air quality monitoring devices are able to gather data on the air quality in various regions of a city since they may be installed to cars [58,59]. Wearable sensors can also track personal exposure to air pollution, providing individuals with information on their exposure and enabling them to make informed decisions about their activities [60].

7.6 AI LPWAN TECHNOLOGIES FOR AIR QUALITY MONITORING

Artificial intelligence LPWAN technologies have become a feasible choice for air quality monitoring due to their capacity to enable long-range, low-power connection [61]. Urban and distant regions may use LPWAN technology to establish wireless sensor networks across wide areas with little infrastructure needed, making them suited for air quality monitoring [16]. One popular LPWAN technology used for air quality monitoring is based on the Long Range (LoRa) protocol. LoRa offers an efficient and low-power communication solution that allows for long-range transmission of sensor data. It operates in the sub-GHz frequency bands, which provide excellent signal propagation characteristics, enabling the transmission of data over several kilometers [62].

A number of sensors are used in an LPWAN-based air quality monitoring system to assess a number of different air quality parameters, including temperature, humidity, PM, CO, NO_2, and O_3 [63]. The data gathered by these sensors is sent through LoRa transceivers to a centralized gateway. The central gateway connects the sensor nodes to the web. After that, the data collected by the sensor nodes is sent either to a distant server located in the cloud or a local data processing unit [64].

The gateway is in charge of handling data aggregation and compression, as well as managing connection with the sensor nodes. It also makes sure that data is sent securely over the AI-enabled LPWAN. One of the critical advantages of LPWAN technologies for air quality monitoring is their low-power consumption [65]. The sensor nodes are designed to operate on battery power for an extended period, allowing for long-term deployments without frequent maintenance or power supply infrastructure [66]. The low-power nature of LPWAN also enables energy harvesting techniques, such as solar panels or kinetic energy harvesters, to power the sensor nodes, further enhancing their autonomy. Real-time monitoring is another essential feature provided by AI-enabled LPWAN-based air quality monitoring systems [67]. Air quality metrics are continually measured by the sensor nodes and sent to the central gateway, allowing for almost real-time monitoring of the situation [68]. This timely data can be utilized for early warning systems, environmental management, and decision-making processes [69].

The collected air quality data can be stored, analyzed, and visualized using cloud-based platforms or local data processing units [70]. Data analysis techniques, such as machine learning algorithms, can be applied to detect patterns, trends, and anomalies in air quality data [71,72]. Visualization tools, including charts, maps, and dashboards, can present the analyzed data user-friendly and intuitively, aiding in data interpretation and decision-making processes. AI-enabled LPWAN-based air quality monitoring systems offer scalability and flexibility [73]. The sensor nodes can be easily deployed and scaled up or down per the monitoring requirements. They can be placed strategically in areas of interest, allowing for comprehensive coverage of the target region. The flexibility of LPWAN technologies also enables the integration of additional sensors or modules to monitor other environmental parameters, expanding the monitoring capabilities beyond air quality [74].

Figure 7.2 visually represents the integration of wireless sensor networks (WSNs) and LPWAN technology, highlighting their role in connecting the

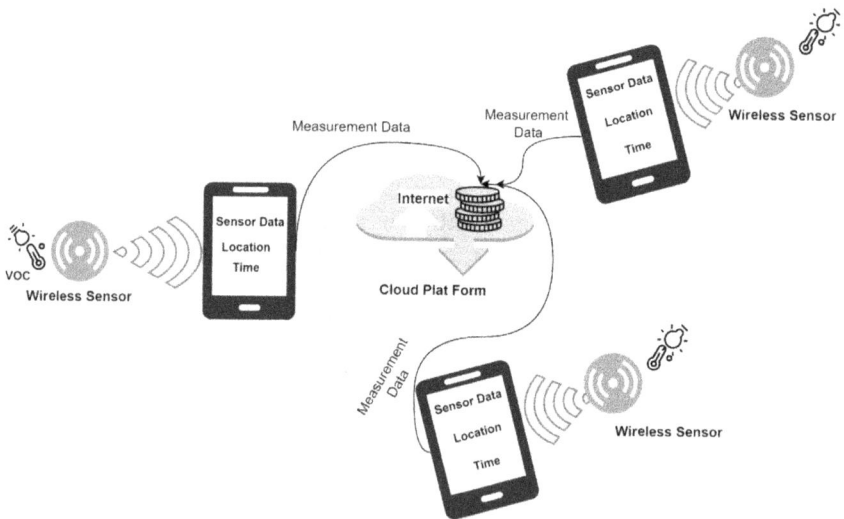

Figure 7.2 Wireless sensor networks and LPWAN technology: connecting the physical world.

physical world. It illustrates the seamless communication and data exchange between sensor devices and the network infrastructure, enabling pervasive monitoring and control of various applications."

7.6.1 Comparison and explanation of LoRaWAN, Sigfox, and NB-IoT

LoRaWAN, Sigfox, and NB-IoT are all LPWAN technologies that can be used for air quality monitoring [75,76]. Here's a look at how these tools compare: LoRaWAN is an open standard LPWAN technology that lets devices communicate over long distances with little power [77]. It uses unlicensed ISM radio bands and can send data over lengths of up to several kilometers [78]. LoRaWAN has a low data rate, usually tens of kilobits per second, but that is enough to send information about air quality [79]. LoRaWAN is also affordable and can work with a lot of nodes. The problem with LoRaWAN is that it costs more up front to set up the infrastructure [80]. Sigfox is an LPWAN technology that is only used by Sigfox. It uses a narrowband radio bandwidth to send data over long distances while using little power [44,81]. It is like LoRaWAN in that it has a low data rate and can connect millions of nodes [82]. Sigfox is cheap and easy to set up because it doesn't need much hardware [83]. But it doesn't cover as much ground and isn't as common as LoRaWAN or NB-IoT. NB-IoT is a cellular LPWAN system that works on an approved frequency. This makes it safer and more reliable [84,85]. It can handle more data than LoRaWAN and Sigfox and thousands of nodes [82,86]. NB-IoT has a higher upfront cost than LoRaWAN or Sigfox, but it offers better coverage and reliability [84,87].

7.6.2 Advantages and disadvantages of each technology

Some advantages and disadvantages of each LPWAN technology for air quality monitoring are shown in Figure 7.3.

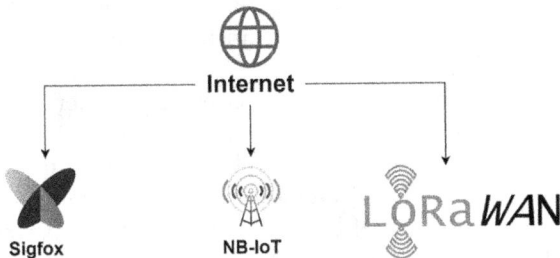

Figure 7.3 Advantages and disadvantages of each technology Sigfox, NBIOT, and LORAWAN.

7.6.2.1 Advantages of LoRaWAN

Low power consumption enables long battery life for devices [88]. Long-range transmission of up to several kilometers makes it ideal for outdoor deployments [89]. Long-range transmission capabilities of LPWAN technologies, such as LoRa, enable communication over several kilometers, making them ideal for outdoor deployments [90]. This feature allows for the placement of sensor nodes in remote areas, industrial sites, and urban environments to monitor air quality conditions. The ability to transmit data over long distances ensures comprehensive coverage and accurate data collection from diverse locations. AI-enabled LPWAN's long-range transmission is particularly beneficial for monitoring air quality in expansive regions or cities [91]. The extended reach minimizes the need for extensive infrastructure deployment, reducing costs and complexity [92].

Supports many nodes, up to millions, on a single network [93]. The scalability of LPWAN allows for the seamless integration of numerous devices within a unified network infrastructure [94]. This capability enables extensive coverage and monitoring across diverse environments, including urban and remote regions [95]. Accommodating many nodes promotes widespread adoption and facilitates the development of comprehensive air quality monitoring systems. The ability to connect millions of nodes on a single network fosters collaboration, innovation, and the exchange of valuable data insights. By supporting a massive scale of deployment, LPWAN technologies empower organizations to gather holistic and accurate air quality data.

Open standard, which enables interoperability and encourages innovation [96], is crucial in fostering and enabling interoperability. Providing a common framework allows different devices and systems to communicate seamlessly. This promotes collaboration among vendors, developers, and researchers, leading to the creation of diverse ecosystems. Open standards eliminate vendor lock-in, ensuring users can choose from compatible products and services [97]. They encourage healthy competition, driving innovation and advancements in technology. The openness and accessibility of open standards create a community of experts who collaborate and contribute to the continuous improvement of the technology. Open standards foster a dynamic and innovative environment, benefiting industries like air quality monitoring [98].

Cost-effective, due to its low infrastructure requirements, with minimal infrastructure, the deployment costs are reduced, making it an economical choice for air quality monitoring systems [99]. The low number of gateways required for coverage further decreases upfront and operational expenses. The low power consumption of LPWAN devices extends battery life, minimizing maintenance costs [100]. Scalability allows for easy expansion without significant infrastructure upgrades or additional expenses.

7.6.2.2 Disadvantages of LoRaWAN

The low data rate, typically tens of kilobits per second, limits the amount of data transmitted; low power consumption in LPWAN devices enables extended battery life, ensuring long-term operation without frequent replacements. This efficiency minimizes maintenance efforts and reduces operational costs [101]. Secondly, the low power requirements of LPWAN technologies contribute to energy-efficient and sustainable air quality monitoring solutions. This energy efficiency reduces operational costs and minimizes frequent battery replacements' environmental impact [102]. By adopting AI-enabled LPWAN-based solutions, organizations can contribute to sustainable practices while effectively monitoring and improving air quality conditions. LPWAN devices can transmit data over long distances by optimizing power consumption while preserving battery life, making them ideal for remote and outdoor deployments. Susceptible to interference from other devices operating on the same frequency band, Limited network coverage in some areas, especially in urban environments with high buildings [103].

7.6.2.3 Advantages of Sigfox

LPWAN technologies offer a range of benefits for air quality monitoring systems. The low power consumption of LPWAN enables devices to have long battery life, ensuring sustained operation without frequent replacements [104]. With its long-range transmission capability of up to several kilometers, LPWAN is well-suited for outdoor deployments, allowing for comprehensive coverage and data collection [105]. Furthermore, the low infrastructure requirements of LPWAN make it easy to deploy, reducing upfront costs and enabling scalability. Overall, LPWAN technologies provide cost-effective and energy-efficient solutions for air quality monitoring, promoting sustainability and effective environmental management [106].

7.6.2.4 Disadvantages of Sigfox

Sigfox, while offering certain advantages, also has some limitations and disadvantages that need to be considered [107]. One of the main drawbacks is its limited data rate and bandwidth. Sigfox operates at a low data rate, typically around 100 bits per second, which restricts the amount of data that can be transmitted [108]. This limitation makes it unsuitable for high-resolution or real-time data streaming applications, such as high-definition video or large data transfers. Another disadvantage of Sigfox is its limited coverage compared to other LPWAN technologies [109]. Sigfox operates on a global network, but its coverage may be limited to specific regions or areas within a country. This restricted coverage can be challenging for applications requiring extensive geographic reach or deployments in remote or rural areas [107].

In addition, Sigfox's subscription-based model can be a disadvantage for some users. Users must pay subscription fees to access the Sigfox network, which may add to the overall cost of deploying and maintaining an air quality monitoring system. This subscription-based approach might not suit organizations, individuals with budget constraints, or those seeking more flexible pricing options [31].

Furthermore, the lack of bidirectional communication is another limitation of Sigfox. Sigfox primarily supports unidirectional communication, meaning that devices can transmit data to the network, but receiving commands or acknowledgments is limited. This limitation restricts the ability to remotely configure or manage devices, potentially requiring physical access for maintenance or updates [110].

7.6.2.5 Advantages of NB-IoT

A wireless communication technique specifically created for IoT devices is called narrowband Internet of Things (NB-IoT) [111]. It offers numerous advantages, making it a preferred choice for various IoT applications [112]. Firstly, NB-IoT provides excellent coverage and penetration capabilities, enabling devices to operate in challenging environments such as basements, underground areas, or remote locations. This technology utilizes a licensed spectrum, ensuring reliable and secure communication, which is crucial for applications requiring sensitive data transmission [113]. NB-IoT devices are energy-efficient and can operate on battery power for several years, minimizing the need for frequent maintenance or battery replacements [114]. NB-IoT is perfect for equipment that needs long-term monitoring or is placed in inhospitable locations because to its low power consumption [114].

Furthermore, NB-IoT offers high scalability, allowing for connecting a massive number of devices within a single network, making it suitable for large-scale deployments [115]. The technology also provides robust interference and noise immunity, ensuring stable and consistent connectivity. Overall, NB-IoT's advantages, including extensive coverage, energy efficiency, security, scalability, and reliability, make it a highly suitable option for many IoT applications [116].

7.6.2.6 Disadvantages of NB-IoT

Due to the many benefits it provides, NB-IoT (Narrowband Internet of Things) technology has recently attracted a lot of interest and gained widespread acceptance within the IoT sector [111]. However, it is essential to acknowledge that NB-IoT also has certain disadvantages. Firstly, one of the fundamental limitations is the relatively low data transmission rate [117].

NB-IoT operates on narrowband frequencies, limiting the quantity of data that can be transmitted at once [118]. This limitation makes it unsuitable for applications that require real-time or high-bandwidth data transfer. Secondly, NB-IoT networks may experience coverage challenges in certain areas, especially in remote or densely populated locations [119]. The technology relies on existing cellular infrastructure, and while efforts have been made to improve coverage, there may still be gaps where devices struggle to connect reliably [120]. Thirdly, the power consumption of NB-IoT devices can be relatively high, especially during data transmission. This can pose challenges for devices with limited power sources, such as battery-operated IoT devices, as it may require more frequent recharging or battery replacement. Lastly, the deployment and maintenance costs associated with NB-IoT can be relatively high, especially for small-scale IoT deployments [87].

These costs include acquiring NB-IoT-enabled devices, integrating them into existing systems, and managing the network infrastructure. Despite these disadvantages, NB-IoT remains a viable technology for various IoT applications, albeit with inevitable trade-offs that need to be considered during implementation [121].

When assessing the effectiveness of an AI LPWAN system for monitoring air quality, range is a key component to take into account, especially if the sensors will be deployed in remote or hard-to-reach areas [122]. LoRaWAN and Sigfox are known for their long-range capabilities, while NB-IoT has a more limited range. Here's a more detailed table summarizing information related to "Wireless-based IoT using AI LPWAN Technology for Quality Air" [31]. Table 7.2 provides a comprehensive overview of various IoT applications that are specifically designed to monitor and measure air quality parameters, such as pollutant levels, particulate matter, and gas concentrations.

7.6.3 Factors to consider when choosing an AI-enabled LPWAN technology for air quality monitoring

The choice of a transmission protocol depends on the requirements of the specific application as well as on how easily the network infrastructure can be accessed [132], When selecting an AI-enabled LPWAN technology for air quality monitoring, several factors need to be taken into consideration as shown in Figure 7.4.

7.6.3.1 Battery life

AI LPWAN technologies are designed to be low-power, meaning the sensors can operate on battery power for an extended period. However, different LPWAN technologies have different power requirements, and battery

Table 7.2 IoT applications related to air quality parameters

AI LPWAN technology	IoT application	Air quality parameters	References
NB-IoT	Smart city monitoring Industrial emissions monitoring Indoor air quality monitoring Traffic-related air pollution monitoring Agricultural air quality monitoring	Particulate Matter (PM) levels CO_2 levels VOCs NO_2 O_3	[112–115,119,123,124]
LoRaWAN	Smart city monitoring	PM Levels	[65]
	Indoor air quality monitoring	CO_2 Level	[125]
	Environmental sensing	VOCs	[126]
	Air quality monitoring for buildings	O_3	[127]
Sigfox	Smart city monitoring	PM Levels	[128]
	Industrial emissions monitoring	CO_2 Level	[129]
	Air quality monitoring for logistics	VOCs	[130]
	Urban green spaces monitoring	NO_2	[131]

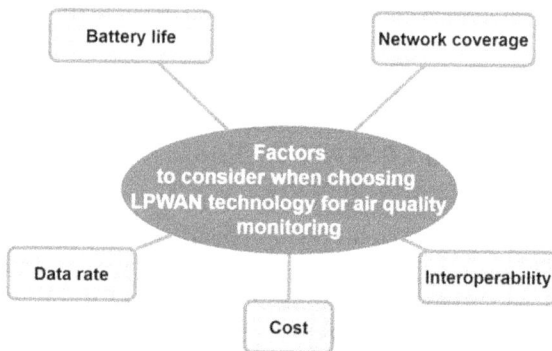

Figure 7.4 Factors to consider when choosing an AI-enabled LPWAN technology for air quality monitoring.

life can vary depending on transmission frequency and data rate. Choosing an LPWAN technology that can provide sufficient battery life for the air quality monitoring system [133,134].

7.6.3.2 Data rate

The needed data rate of the AI LPWAN technology will depend on how much data the air quality monitoring system has to broadcast. Due to their lower data rates than NB-IoT, LoRaWAN and Sigfox may not be appropriate for applications requiring the transmission of huge volumes of data [134].

7.6.3.3 Network coverage

Network coverage availability is an essential factor to consider when choosing an AI LPWAN technology. LoRaWAN and Sigfox networks are typically deployed by third-party providers, which can limit their coverage in some areas. In contrast, NB-IoT uses existing cellular infrastructure, meaning it has broader network coverage [134,135].

7.6.3.4 Interoperability

Interoperability is essential for integrating air quality monitoring systems with other devices or systems. LoRaWAN is an open standard, meaning it is more interoperable than Sigfox, which is proprietary. NB-IoT is also based on standard cellular technology, which makes it more interoperable with other cellular devices [136,137].

7.6.3.5 Cost

Finally, the cost is essential when choosing an AI LPWAN technology for air quality monitoring. LoRaWAN and Sigfox have lower infrastructure costs than NB-IoT, which requires a cellular network infrastructure. However, the sensors' cost can vary depending on the LPWAN technology [51].

7.7 DEPLOYING AI-ENABLED LPWAN SENSORS

7.7.1 Strategic placement of sensors in areas with high pollution levels

Strategic placement of sensors in areas with high pollution levels is an important consideration when deploying air quality monitoring systems [138]. Several factors should be taken into account when deciding where to place sensors, including.

7.7.1.1 Proximity to pollution sources

Sensors should be placed close to sources of pollution, such as factories, highways, or industrial zones, to capture the level of pollutants in the area accurately. By placing sensors close to the source, it is possible to identify the specific types of pollutants and their concentration levels [139].

7.7.1.2 Airflow patterns

Airflow patterns play an essential role in determining where to place sensors. Wind and other weather patterns can carry air pollution over long distances. So it is essential to consider the direction and speed of airflow when determining sensor placement [140].

7.7.1.3 Accessibility

Sensors should be placed in areas that are easily accessible for maintenance and battery replacement. This is especially important in areas with high pollution levels, as the sensors may need to be checked and maintained more frequently to ensure accuracy [141].

7.7.1.4 Variability in pollution levels

Pollution levels can vary significantly within a given area. So it is essential to place sensors in locations that accurately capture the range of pollution levels in the area being monitored. This may involve placing sensors in multiple locations within the same area to capture the variability in pollution levels [142].

7.7.1.5 Integration with other monitoring systems

Locations where sensors can be integrated with other monitoring systems, such as weather stations or traffic monitoring systems, should be chosen. This may provide a more complete picture of the environmental circumstances in the region being watched and assist in locating possible pollution sources [143].

Strategically placing sensors in areas with high pollution levels is crucial for accurate and effective air quality monitoring. By considering factors such as proximity to pollution sources, airflow patterns, accessibility, variability in pollution levels, and integration with other monitoring systems, it is feasible to create a monitoring system that offers insightful data on the levels of pollution and air toxicity in the monitored region. Types of air pollutants that can be monitored using AI-enabled sensorsAI Enabled LPWAN sensors can be used to monitor a wide range of air pollutants, including [144].

7.7.1.6 Particulate Matter

Particulate matter (PM) is a conglomeration of very small particles and droplets that may be breathed in and lead to a variety of respiratory issues [144]. Most routinely measured are PM2.5 and PM10. PM2.5 particles are 2.5 µm or less, whereas PM10 particles are 10 µm or less. AI-enabled LPWAN sensors can measure PM2.5 and PM10 levels in the air.

7.7.1.7 Nitrogen Oxides (NOx)

NOx is a collection of very reactive gases that may lead to respiratory issues, pollution, and the creation of acid rain. NOx is often created by industrial processes and automobile emissions. LPWAN sensors with AI-enabled may be used to track the amount of NOx in the air [27,145].

7.7.1.8 Sulfur Dioxide (SO₂)

SO_2 is a gas produced by the combustion of fossil fuels that can cause respiratory issues and contribute to acid rain formation. AI-enabled LPWAN sensors can be used to monitor SO_2 levels in the atmosphere [146].

7.7.1.9 Carbon Monoxide (CO)

Headaches, dizziness, and nausea may result from the incomplete combustion of fossil fuels, which produces the gas known as CO. The level of CO in the air may be monitored using sensors that are AI-enabled [40].

7.7.1.10 Ozone (O₃)

Nitrogen oxides and volatile organic molecules react under sunlight to produce the gas O_3. In addition to contributing to smog production, it may result in respiratory issues. LPWAN sensors with AI enabled may be used to track the amount of O_3 in the air [146].

7.7.1.11 Volatile organic compounds (VOCs)

VOCs are organic substances that may evaporate into the atmosphere and cause smog to occur. They may also impact other aspects of health, such as respiratory issues. VOC concentrations in the air may be monitored using AI-enabled sensors. Overall, a variety of air contaminants may be tracked using AI-enabled LPWAN sensors, providing useful information about the degree of pollution and air quality in a specific location. Monitoring these chemicals makes it easier to identify probable pollution sources, decrease exposure, and take action to enhance air quality [146,147].

7.7.2 Calibration and maintenance of sensors

Calibration and maintenance of sensors are essential to ensuring accurate and reliable readings. Here are some critical aspects of calibration and maintenance for AI-enabled LPWAN sensors [16].

7.7.2.1 Calibration

Calibration is adjusting the sensor readings to match the actual values of the pollutant concentration in the air. This involves comparing the sensor readings with a reference standard and adjusting the readings as needed. Calibration should be done regularly, especially when the sensor has been moved or after a significant period of use [148].

7.7.2.2 Maintenance

Regular maintenance ensures the sensor functions appropriately and provides accurate readings. Maintenance may include cleaning the sensor, replacing the battery or other components, and checking for any physical damage or wear and tear [149].

7.7.2.3 Quality control

Quality control measures should be in place to ensure that the sensors are providing accurate and reliable readings. This may involve regular checks of the sensor data, comparison with other monitoring systems, and cross-validation of data from multiple sensors [150].

7.7.2.4 Data management

Data management is essential to sensor maintenance, which involves collecting, storing, and analyzing sensor data. Data should be collected at regular intervals and stored securely. Data analysis should be done regularly to identify trends and anomalies and make necessary adjustments to the monitoring system [151].

7.7.2.5 Record keeping

It is essential to maintain records of sensor calibration, maintenance, and quality control measures. This helps ensure that the sensors are correctly maintained and calibrated and can be used to identify any issues that may arise over time. In summary, calibration and maintenance of sensors are essential to ensure accurate and reliable readings. Regular calibration and maintenance, quality control measures, data management, and record-keeping are essential aspects of maintaining a robust and effective air quality monitoring system [152].

7.8 DATA TRANSMISSION AND MANAGEMENT

Data transmission and management are critical components of any AI-enabled LPWAN sensor network, as they ensure that the data collected by the sensors is transmitted reliably and efficiently and can be analyzed and acted upon promptly. Here are some critical aspects of data transmission and management for AI-enabled LPWAN sensors:

7.8.1 Data transmission

AI-enabled LPWAN sensors use low-power, wide-area networks to transmit data over long distances. These networks use technologies like LoRaWAN, Sigfox, and NB-IoT. The sensors may transmit data regularly or respond to specific events or triggers. It is essential to ensure that the data transmission is reliable and that the sensor data is received and processed correctly [153].

7.8.2 Data management

Once the sensor data has been transmitted, it must be stored, processed, and analyzed. This may include processing the data locally on the sensor or gateway device using edge computing or storing and processing it on a cloud-based platform. To make analysis and display of the data quick and simple, it should be arranged in a systematic and uniform style [154].

7.8.3 Data security

When sending and processing sensor data, data security is a crucial factor. To secure the data in transit, LPWAN networks use encryption and other security measures. However, it is also crucial to make sure that the data is maintained securely and that only authorized individuals have access to it [48,155].

7.8.4 Data visualization

Sensor data can be visualized using various tools, such as dashboards and maps, to enable stakeholders to understand the data and take action. Data visualization can help identify patterns, trends, and anomalies and can be used to clearly and concisely communicate the data to stakeholders [139].

7.8.5 Data sharing

Sharing sensor data with other stakeholders, such as government agencies or research institutions, can help to improve the understanding of air quality issues and inform policy decisions. However, it is essential to ensure that data sharing is done securely and responsibly and that the privacy of individuals is protected [155].

Data transmission and management are critical components of an AI-enabled LPWAN sensor network. Reliable data transmission, secure data storage and management, data visualization, and responsible data sharing are all critical considerations in ensuring the sensor data is used effectively to improve air quality and inform decision-making.

7.8.6 Wireless transmission of data from sensors to a central server

Wireless data transmission from sensors to a central server is a crucial aspect of AI-enabled LPWAN-sensor networks. Here are some essential considerations for wireless data transmission [156]:

7.8.7 Network architecture

The LPWAN network architecture plays a critical role in wireless data transmission. The network may be based on a star topology, where sensors communicate directly with a central gateway, or a mesh topology, where sensors communicate with multiple gateways or other sensors to ensure redundancy and reliability [157].

7.8.8 Frequency band

LPWAN networks operate in various frequency bands, such as the unlicensed ISM band, which LoRaWAN and Sigfox use. It is vital to ensure that the frequency band the network uses is compatible with the region's regulatory requirements and the specific application [158].

7.8.9 Transmission protocol

LPWAN networks use transmission protocols, such as LoRaWAN, Sigfox, and NB-IoT. Each protocol has strengths and weaknesses, such as range, data rate, and power consumption. The selection of a transmission protocol is dependent on the particular application needs as well as the accessibility of the network infrastructure [159].

7.8.10 Security

Hacking and other security risks have been linked to wireless data transfer. Data in transit is encrypted and otherwise protected on LPWAN networks. It is still important to keep the data in a safe place where only authorized people can access it [160].

7.8.11 Data throughput

LPWAN networks have limited data throughput, which may be a constraint in specific applications. It is essential to ensure the data rate is sufficient to meet the application requirements while optimizing power consumption and network resources [161].

Wireless data transmission from sensors to a central server is a critical component of AI-enabled LPWAN sensor networks. The network architecture, frequency band, transmission protocol, security, and data throughput are all essential to ensure reliable and efficient data transmission.

7.8.12 Real-time monitoring and visualization of air quality data

Real-time monitoring and visualization of air quality data refers to the continuous collection and analysis of air quality data from AI-enabled LPWAN sensors and the real-time display of that data in a visual format for stakeholders to view [162]. This enables stakeholders to quickly identify changes in air quality and take appropriate actions to mitigate any adverse effects. Real-time monitoring and visualization of air quality data are essential because they provide stakeholders with up-to-date information on air quality conditions. This data may be used to locate polluted locations, observe trends over time, and evaluate the efficiency of air quality management efforts. By accessing this information, stakeholders can make informed decisions about the actions they need to take to protect public health and the environment [163].

Data visualization is critical in real-time monitoring and visualization of air quality data. It enables stakeholders to see air quality data clearly and concisely using visual tools such as maps, charts, and graphs. Visualization allows stakeholders to identify areas with high pollution levels quickly and can help identify trends and patterns in air quality data over time. The alert system can be set up to notify stakeholders when air quality conditions reach a certain threshold, such as when the concentration of a particular pollutant exceeds a certain level. This can help trigger appropriate actions, such as issuing health advisories or implementing air quality control measures [164].

Historical data analysis is another crucial aspect of real-time monitoring and visualization of air quality data. By analyzing historical data, stakeholders can identify long-term trends and patterns in air quality data, which can help inform decisions about air quality control measures. Historical data analysis can also be used to assess the effectiveness of air quality control measures over time. In summary, real-time monitoring and visualization of air quality data are critical in AI-enabled LPWAN sensor networks. It allows interested parties to spot changes in air quality rapidly, monitor trends over time, and take the necessary precautions to safeguard environmental and human health. Data visualization, alert systems, and historical data analysis are essential in ensuring stakeholders have the information to make informed decisions and take appropriate actions to improve air quality. Integration with other systems, such as weather monitoring and traffic management [165].

Integrating AI-enabled LPWAN air quality sensor networks with other systems, such as weather monitoring and traffic management, can provide additional insights into the factors affecting air quality and enable more effective decision-making. Weather monitoring systems can provide data on temperature, humidity, and wind speed, which can be used to better understand how weather conditions impact air quality. For example, high temperatures and low wind speeds can increase the concentration of pollutants in the air, while rain and strong winds can help disperse pollutants. Integrating weather monitoring data with air quality data from AI-enabled LPWAN sensors can help stakeholders identify the relationships between weather conditions and air quality and take appropriate actions to mitigate any adverse effects [166].

Traffic management systems can also be integrated with AI-enabled LPWAN-air quality sensor networks. In metropolitan areas, transportation is a key source of air pollution. By combining traffic and air quality data, stakeholders may find places where there is a lot of pollution from vehicles. This information can inform decisions about traffic management strategies, such as implementing traffic calming measures or promoting the use of public transport. Integrating air quality sensor networks with other systems can also enable the development of more advanced predictive models for air quality. By combining data from multiple sources, such as air quality sensors, weather monitoring systems, and traffic management systems, stakeholders can develop models that predict future air quality based on current and forecasted conditions. This can help to inform decisions about air quality control measures, such as issuing health advisories or implementing traffic management strategies [48,167].

In summary, integrating AI-enabled LPWAN air quality sensor networks with other systems, such as weather monitoring and traffic management, can provide valuable insights into the factors that impact air quality and enable more effective decision-making. By combining data from multiple sources, stakeholders can develop more advanced models for predicting future air quality conditions and inform decisions about air quality control measures.

7.9 BENEFITS OF AI-ENABLED LPWAN-BASED IoT NETWORKS FOR AIR QUALITY MONITORING

Low-Power Wide-Area Network (AI-enabled LPWAN) is a wireless network that enables long-range and low-power communication for IoT devices. AI-enabled LPWAN-based IoT networks can be used for air quality monitoring in several ways, providing several benefits over traditional monitoring methods [168,169].

7.9.1 Cost-effective

AI-enabled LPWAN-based IoT networks are cost-effective and can be easily deployed over large areas. This makes deploying many air quality sensors across a city or region possible, providing more accurate and detailed information about air quality levels [169].

7.9.2 Low-power consumption

IoT devices with AI capabilities that run on LPWAN need extremely little power, so they may last for years on a single battery. This makes them ideal for outdoor air quality monitoring, where access to power sources may be limited [169].

7.9.3 Long-range connectivity

AI-enabled LPWAN-based IoT devices can communicate over long distances, even in areas with poor cellular network coverage. This makes it possible to monitor air quality in remote areas, which are often more vulnerable to air pollution [65].

7.9.4 Real-time data monitoring

AI-enabled LPWAN-based IoT devices provide real-time data monitoring, allowing users to track air quality changes. This helps authorities to take immediate action to address air pollution issues [65].

7.9.5 Easy to install and maintain

AI-enabled LPWAN-based IoT devices are easy to install and maintain, as they require minimal infrastructure and can be easily connected to the internet. This means that air quality monitoring can be implemented quickly and cost-effectively. Overall, AI-enabled LPWAN-based IoT networks offer a cost-effective and reliable solution for air quality monitoring, providing real-time data monitoring, long-range connectivity, and low power consumption. They are an excellent option for urban and rural locations where air quality monitoring is necessary for ensuring the public's health and safety because of these advantages [65].

7.10 CASE STUDIES AND EXAMPLES

Smart City Air Challenge, Madrid: In 2018, the city of Madrid launched the Smart City Air Challenge, which involved deploying a network of AI-enabled LPWAN-based IoT sensors to monitor air quality in the city. The sensors were installed on streetlights and public buildings, providing real-time data on air quality levels. The data was made available to the public through a mobile app, enabling citizens to make informed decisions about their daily activities.

7.10.1 Clean Air Belgium, Brussels

In 2020, Clean Air Belgium deployed a network of AI-enabled LPWAN-based IoT sensors to monitor air quality in Brussels. The sensors were installed on lamp posts and public buildings across the city, providing real-time data on air pollution levels. The data was made available to the public through a web-based dashboard, enabling citizens to track air quality levels in their neighborhoods [170].

7.10.2 AirQo, Uganda

In 2019, the AirQo project deployed a network of AI-enabled LPWAN-based IoT sensors to monitor air quality in Kampala, Uganda. The sensors were installed on rooftops and public buildings across the city, providing real-time data on air quality levels. The data was used to identify pollution hotspots and inform policy decisions to reduce air pollution levels [171].

7.10.3 Eddy, California

In 2020, the Eddy project deployed a network of AI-enabled LPWAN-based IoT sensors to monitor air quality in the Central Valley region of California. The sensors were installed on light poles and regional public buildings, providing real-time data on air pollution levels. The data was used to inform public health and safety decisions, particularly for vulnerable populations like children and older people [172].

7.11 PROSPECTS AND CHALLENGES

7.11.1 Emerging AI LPWAN technologies and their potential for air quality monitoring

New LPWAN technologies portend well for the development of AI-powered IoT networks for air quality monitoring. The Narrowband Internet of Things (NB-IoT), also known as 5G AI-enabled LPWAN, offers greater bandwidth and lower latency than current LPWAN technology [6]. It can increase the accuracy and speed of air quality monitoring, and it works well in high-density metropolitan regions and industrial settings [173]. LTE-M is a low-power, wide-area network technology that improves upon standard cellular networks in terms of capacity and latency. It can offer more widespread coverage for air quality monitoring and is hence well-suited for low-density urban and rural locations [174].

Providing long-distance connection and minimal power consumption, LoRaWAN is a low-power, wide-area network technology. It is appropriate for air quality monitoring in far-off places with inadequate network coverage [175]. A low-power, wide-area network technology called Sigfox provides long-distance communication while using little power. It is appropriate for monitoring air quality in rural places with inadequate network connectivity [12]. Implementing AI-enabled LPWAN-based IoT networks for air quality monitoring still presents certain difficulties, however. Air quality monitoring is often governed by complicated legal frameworks, making the deployment of AI-enabled LPWAN-based IoT networks in certain areas hard [176]. Massive volumes of data are produced by AI-enabled LPWAN-based IoT networks, which may be difficult to manage and analyze. To guarantee the data is handled and analyzed efficiently, solid data management solutions must be in place [177]. To properly integrate and analyze the data obtained from various sensors, compatibility across various LPWAN technologies is required [44].

AI-enabled LPWAN-based IoT sensors require low-power consumption to ensure the batteries last long periods without replacement. There is a need for continued research into developing more energy-efficient sensors.

7.11.2 Challenges in scaling up AI-enabled LPWAN-based IoT networks for widespread air quality monitoring

Scaling up AI-enabled LPWAN-based IoT networks for widespread air quality monitoring presents several challenges. Deploying an AI-enabled LPWAN-based IoT network for air quality monitoring on a large scale can be costly, particularly in regions with large populations. Costs associated with hardware, maintenance, data storage, and data analysis can add up quickly [178]. AI-enabled LPWAN-based IoT networks have limited network coverage, making it challenging to scale to cover large regions. Extending network coverage can be difficult in rural and remote areas that lack infrastructure [84]. Scaling up AI-enabled LPWAN-based IoT networks for air quality monitoring generates large amounts of data. Processing, storing, and analyzing this data requires a robust data management system that can handle large data volumes in real-time. AI-enabled LPWAN-based IoT networks collect and transmit sensitive data about citizens, which can raise privacy and security concerns [44]. Ensuring data privacy and security becomes more challenging as the network scales up to cover large populations. Interoperability between AI-enabled LPWAN-based IoT networks is essential for scaling up air quality monitoring. This allows data to be integrated and analyzed effectively, regardless of the network used to collect it. AI-enabled LPWAN-based IoT networks must comply with data collection, storage, and sharing regulations for air quality monitoring. Meeting regulatory requirements can be challenging when scaling the network to cover a large population [179].

Addressing these challenges requires collaboration between stakeholders, including governments, private companies, and communities. Governments can invest in building infrastructure and provide funding to support the deployment and scaling up of AI-enabled LPWAN-based IoT networks. Private companies can develop innovative solutions that address the challenges of scaling up air quality monitoring networks. Communities can provide feedback and participate in deploying air quality monitoring networks, helping build trust and encourage adoption [179].

7.12 CONCLUSION

In conclusion, wireless-based IoT systems utilizing Low-Power Wide-Area Network (AI-enabled LPWAN) technologies such as LoRaWAN, Sigfox, and NB-IoT offer promising solutions for monitoring and maintaining air quality. These LPWAN technologies are ideal for deploying IoT devices over a wide area to collect data on air quality because they provide efficient, affordable, and long-range communication capabilities. LoRaWAN, with its long-range capability and low power usage, enables the deployment of numerous air quality sensors in a large-scale network. It allows for low-cost infrastructure and offers excellent coverage, making it suitable for monitoring air quality in remote or difficult-to-access locations. The scalability of LoRaWAN networks ensures that additional sensors can be easily integrated as the demand for monitoring expands.

Sigfox, another LPWAN technology, also proves beneficial for air quality monitoring. It operates in unlicensed spectrum bands, providing comprehensive area coverage while requiring low energy consumption. Sigfox's network architecture is highly optimized for low-cost, low-bandwidth applications, making it an attractive choice for applications requiring minimal data transmission, such as transmitting air quality measurements periodically. NB-IoT, a cellular LPWAN technology, leverages existing cellular infrastructure to provide widespread coverage for IoT devices. It offers higher data rates than LoRaWAN and Sigfox, enabling more detailed and frequent air-quality data transmission. NB-IoT's integration with existing cellular networks makes it a convenient choice for urban environments or areas already covered by cellular infrastructure.

Several advantages are achieved by utilizing LPWAN technologies like LoRaWAN, Sigfox, and NB-IoT for air quality monitoring. These technologies enable real-time monitoring of air quality parameters such as PM, gas concentrations, and environmental conditions, allowing for early detection of pollution events or anomalies. The gathered data may be utilized for thorough analysis and modeling to pinpoint pollution sources, comprehend trends, and create successful air quality improvement plans. Moreover, AI-enabled LPWAN-based IoT systems offer cost advantages due to their low power consumption, extended battery life, and reduced infrastructure requirements. The long-range capabilities of these technologies allow for large-scale deployments, ensuring extensive coverage and data collection from various locations simultaneously. This extensive coverage and data collection create a comprehensive and accurate understanding of regional or global air quality.

The integration of AI and low-powered wireless networks has the potential to revolutionize air quality monitoring and management. By harnessing the power of AI algorithms and leveraging the scalability of low-powered wireless networks, we can create a sustainable and efficient framework for ensuring quality air for all. Continued research and development in this field are crucial for building healthier and more environmentally conscious communities in the future.

REFERENCES

[1] P. Fremantle, "A reference architecture for the internet of things," *WSO2 White Pap.*, pp. 2–4, 2015.

[2] B. Di Martino, M. Rak, M. Ficco, A. Esposito, S. A. Maisto, and S. Nacchia, "Internet of things reference architectures, security and interoperability: A survey," *Internet of Things*, vol. 1, pp. 99–112, 2018.

[3] K. K. Patel, S. M. Patel, and P. Scholar, "Internet of things-IOT: Definition, characteristics, architecture, enabling technologies, application & future challenges," *Int. J. Eng. Sci. Comput.*, vol. 6, no. 5, pp. 6122–6131, 2016.

[4] M. Ibrahim, A. Elgamri, S. Babiker, and A. Mohamed, "Internet of things based smart environmental monitoring using the Raspberry-Pi computer," In *2015 Fifth International Conference on Digital Information Processing and Communications (ICDIPC)*, IEEE, 2015, pp. 159–164.

[5] J. Petäjäjärvi, K. Mikhaylov, R. Yasmin, M. Hämäläinen, and J. Iinatti, "Evaluation of LoRa LPWAN technology for indoor remote health and well-being monitoring," *Int. J. Wirel. Inf. Networks*, vol. 24, pp. 153–165, 2017.

[6] I. U. Khan, A. Abdollahi, A. Jamil, B. Baig, M. A. Aziz, and F. Subhan, "A novel design of FANET routing protocol aided 5G communication using IoT," *J. Mob. Multimed.*, vol. 18, no. 5 pp. 1333–1354, 2022.

[7] Z. E. Ahmed, R. A. Saeed, A. Mukherjee, and S. N. Ghorpade, "Energy optimization in low-power wide area networks by using heuristic techniques," *LPWAN Technol. IoT M2M Appl.*, pp. 199–223, 2020.

[8] J. P. Queralta, T. N. Gia, H. Tenhunen, and T. Westerlund, "Edge-AI in LoRa-based health monitoring: Fall detection system with fog computing and LSTM recurrent neural networks," In *2019 42nd International Conference on Telecommunications and Signal Processing (TSP)*, IEEE, 2019, pp. 601–604.

[9] X. Jing, X. Tian, and C. Du, "LPAI-A complete AIoT framework based on LPWAN applicable to acoustic scene classification scenarios," *Sensors*, vol. 22, no. 23, p. 9404, 2022.

[10] M. Usama et al., "Predictive modelling of compression strength of waste GP/FA blended expansive soils using multi-expression programming," *Constr. Build. Mater.*, vol. 392, p. 131956, 2023.

[11] X. Feng, F. Yan, and X. Liu, "Study of wireless communication technologies on Internet of Things for precision agriculture," *Wirel. Pers. Commun.*, vol. 108, no. 3, pp. 1785–1802, Oct. 2019. doi: 10.1007/s11277-019-06496-7.

[12] M. Bembe, A. Abu-Mahfouz, M. Masonta, and T. Ngqondi, "A survey on low-power wide area networks for IoT applications," *Telecommun. Syst.*, vol. 71, pp. 249–274, 2019.

[13] C. A. Trasviña-Moreno, R. Blasco, R. Casas, and A. Asensio, "A network performance analysis of LoRa modulation for LPWAN sensor devices," In *Ubiquitous Computing and Ambient Intelligence: 10th International Conference, UCAmI 2016, San Bartolomé de Tirajana, Gran Canaria*, Spain, *Part II 10*, Springer, November 29–December 2, 2016, pp. 174–181.

[14] M. Iqbal, A. Y. M. Abdullah, and F. Shabnam, "An application based comparative study of LPWAN technologies for IoT environment," In *2020 IEEE Region 10 Symposium (TENSYMP)*, IEEE, 2020, pp. 1857–1860. doi: 10.1109/TENSYMP50017.2020.9230597.

[15] H. Zhu et al., "Index of low-power wide area networks: A ranking solution toward best practice," *IEEE Commun. Mag.*, vol. 59, no. 4, pp. 139–144, 2021.

[16] K. Zheng, S. Zhao, Z. Yang, X. Xiong, and W. Xiang, "Design and implementation of LPWA-based air quality monitoring system," *IEEE Access*, vol. 4, pp. 3238–3245, 2016.

[17] R. Firdaus, M. A. Murti, and I. Alinursafa, "Air quality monitoring system based internet of things (IoT) using LPWAN LoRa," In *2019 IEEE international conference on internet of things and intelligence system (IoTaIS)*, IEEE, 2019, pp. 195–200.

[18] A. Simo, S. Dzitac, I. Dzitac, M. Frigura-Iliasa, and F. M. Frigura-Iliasa, "Air quality assessment system based on self-driven drone and LoRaWAN network," *Comput. Commun.*, vol. 175, pp. 13–24, 2021.

[19] I. U. Khan *et al.*, "Intelligent detection system enabled attack probability using Markov chain in aerial networks," *Wirel. Commun. Mob. Comput.*, vol. 2021, pp. 1–9, 2021.

[20] K. M. Simitha and M. S. S. Raj, "IoT and WSN based air quality monitoring and energy saving system in SmartCity project," In *2019 2nd International Conference on Intelligent Computing, Instrumentation and Control Technologies (ICICICT)*, IEEE, 2019, pp. 1431–1437.

[21] A. Zanella, N. Bui, A. Castellani, L. Vangelista, and M. Zorzi, "Internet of things for smart cities," *IEEE Internet Things J.*, vol. 1, no. 1, pp. 22–32, 2014.

[22] Y. Mehmood, F. Ahmad, I. Yaqoob, A. Adnane, M. Imran, and S. Guizani, "Internet-of-things-based smart cities: Recent advances and challenges," *IEEE Commun. Mag.*, vol. 55, no. 9, pp. 16–24, 2017.

[23] I. U. Khan, I. M. Qureshi, M. A. Aziz, T. A. Cheema, and S. B. H. Shah, "Smart IoT control-based nature inspired energy efficient routing protocol for flying ad hoc network (FANET)," *IEEE Access*, vol. 8, pp. 56371–56378, 2020.

[24] M. Ahsan, M. A. Based, J. Haider, and E. M. G. Rodrigues, "Smart monitoring and controlling of appliances using LoRa based IoT system," *Designs*, vol. 5, no. 1, p. 17, 2021.

[25] S.-Y. Wang, J.-J. Zou, Y.-R. Chen, C.-C. Hsu, Y.-H. Cheng, and C.-H. Chang, "Long-term performance studies of a LoRaWAN-based PM2.5 application on campus," In *2018 IEEE 87th Vehicular Technology Conference (VTC Spring)*, IEEE, 2018, pp. 1–5.

[26] Y. Zhang, F. Ren, A. Wu, T. Zhang, J. Cao, and D. Zheng, "Certificateless multi-party authenticated encryption for NB-IoT terminals in 5G networks," *IEEE Access*, vol. 7, pp. 114721–114730, 2019.

[27] R. K. Singh, M. Aernouts, M. De Meyer, M. Weyn, and R. Berkvens, "Leveraging LoRaWAN technology for precision agriculture in greenhouses," *Sensors*, vol. 20, no. 7, p. 1827, 2020.

[28] M. Shahjalal, M. K. Hasan, M. M. Islam, M. M. Alam, M. F. Ahmed, and Y. M. Jang, "An overview of AI-enabled remote smart-home monitoring system using LoRa," In *2020 International Conference on Artificial Intelligence in Information and Communication (ICAIIC)*, IEEE, 2020, pp. 510–513.

[29] C. Jiang, Y. Yang, X. Chen, J. Liao, W. Song, and X. Zhang, "A new-dynamic adaptive data rate algorithm of LoRaWAN in harsh environment," *IEEE Internet Things J.*, vol. 9, no. 11, pp. 8989–9001, 2021.

[30] J. P. Queralta, T. N. Gia, Z. Zou, H. Tenhunen, and T. Westerlund, "Comparative study of LPWAN technologies on unlicensed bands for M2M communication in the IoT: Beyond LoRa and LoRaWAN," *Procedia Comput. Sci.*, vol. 155, pp. 343–350, 2019.

[31] K. Mekki, E. Bajic, F. Chaxel, and F. Meyer, "Overview of cellular LPWAN technologies for IoT deployment: Sigfox, LoRaWAN, and NB-IoT," In *2018 IEEE International Conference on Pervasive Computing and Communications Workshops (Percom Workshops)*, IEEE, 2018, pp. 197–202.

[32] M. Aernouts, R. Berkvens, K. Van Vlaenderen, and M. Weyn, "Sigfox and LoRaWAN datasets for fingerprint localization in large urban and rural areas," *Data*, vol. 3, no. 2, p. 13, 2018.

[33] I. U. Khan, S. B. H. Shah, L. Wang, M. A. Aziz, T. Stephan, and N. Kumar, "Routing protocols & unmanned aerial vehicles autonomous localization in flying networks," *Int. J. Commun. Syst.*, p. e4885, 2021.

[34] D. K. Choubey, V. Shukla, V. Soni, J. Kumar, and D. K. Dheer, "A review on IoT architectures, protocols, security, and applications," *Ind. Internet Things*, pp. 225–242, 2022.

[35] Q. Wang and P. Lu, "Research on application of artificial intelligence in computer network technology," *Int. J. Pattern Recognit. Artif. Intell.*, vol. 33, no. 05, p. 1959015, 2019.

[36] D. Lee and S. N. Yoon, "Application of artificial intelligence-based technologies in the healthcare industry: Opportunities and challenges," *Int. J. Environ. Res. Public Health*, vol. 18, no. 1, p. 271, 2021.

[37] H. Farzaneh, L. Malehmirchegini, A. Bejan, T. Afolabi, A. Mulumba, and P. P. Daka, "Artificial intelligence evolution in smart buildings for energy efficiency," *Appl. Sci.*, vol. 11, no. 2, p. 763, 2021.

[38] A. Abdollahi and M. Fathi, "An intrusion detection system on ping of death attacks in IoT networks," *Wirel. Pers. Commun.*, vol. 112, pp. 2057–2070, 2020.

[39] E. U. Ogbodo, A. M. Abu-Mahfouz, and A. M. Kurien, "A Survey on 5G and LPWAN-IoT for improved smart cities and remote area applications: From the aspect of architecture and security," *Sensors*, vol. 22, no. 16, p. 6313, 2022.

[40] S. Duangsuwan, A. Takarn, R. Nujankaew, and P. Jamjareegulgarn, "A study of air pollution smart sensors lpwan via nb-iot for thailand smart cities 4.0," In *2018 10th International conference on knowledge and smart technology (KST)*, IEEE, 2018, pp. 206–209.

[41] N. Tsavalos and A. Abu Hashem, "Low power wide area network (LPWAN) Technologies for Industrial IoT applications," 2018.

[42] N. Nurelmadina *et al.*, "A systematic review on cognitive radio in low power wide area network for industrial IoT applications," *Sustainability*, vol. 13, no. 1, p. 338, 2021.

[43] M. L. Liya and D. Arjun, "A survey of LPWAN technology in agricultural field," In *2020 Fourth International Conference on I-SMAC (IoT in Social, Mobile, Analytics and Cloud)(I-SMAC)*, IEEE, 2020, pp. 313–317.

[44] B. S. Chaudhari, M. Zennaro, and S. Borkar, "LPWAN technologies: Emerging application characteristics, requirements, and design considerations," *Futur. Internet*, vol. 12, no. 3, p. 46, Mar. 2020. doi: 10.3390/fi12030046.

[45] F. Lubrano, D. Sergi, F. Bertone, and O. Terzo, "Multi-network technology cloud-based asset-tracking platform for IoT devices," In *Complex, Intelligent and Software Intensive Systems: Proceedings of the 14th International Conference on Complex, Intelligent and Software Intensive Systems (CISIS-2020)*, Springer, 2021, pp. 344–354.

[46] Y. Chen and D. Han, "Water quality monitoring in smart city: A pilot project," *Autom. Constr.*, vol. 89, pp. 307–316, 2018.

[47] K. Tzortzakis, K. Papafotis, and P. P. Sotiriadis, "Wireless self powered environmental monitoring system for smart cities based on LoRa," In *2017 Panhellenic Conference on Electronics and Telecommunications (PACET)*, IEEE, 2017, pp. 1–4.

[48] D. D. Olatinwo, A. Abu-Mahfouz, and G. Hancke, "A survey on LPWAN technologies in WBAN for remote health-care monitoring," *Sensors*, vol. 19, no. 23, p. 5268, 2019.

[49] M. M. Alam, H. Malik, M. I. Khan, T. Pardy, A. Kuusik, and Y. Le Moullec, "A survey on the roles of communication technologies in IoT-based personalized healthcare applications," *IEEE Access*, vol. 6, pp. 36611–36631, 2018.

[50] L. Germani, V. Mecarelli, G. Baruffa, L. Rugini, and F. Frescura, "An IoT architecture for continuous livestock monitoring using LoRa LPWAN," *Electronics*, vol. 8, no. 12, p. 1435, 2019.

[51] K. Mekki, E. Bajic, F. Chaxel, and F. Meyer, "A comparative study of LPWAN technologies for large-scale IoT deployment," *ICT Express*, vol. 5, no. 1, pp. 1–7, 2019.

[52] S. Jain, M. Pradish, A. Paventhan, M. Saravanan, and A. Das, "Smart energy metering using LPWAN IoT technology," In *ISGW 2017: Compendium of Technical Papers: 3rd International Conference and Exhibition on Smart Grids and Smart Cities*, Springer, 2018, pp. 19–28.

[53] Y. Kabalcı and M. Ali, "Emerging LPWAN technologies for smart environments: An outlook," In *2019 1st Global Power, Energy and Communication Conference (GPECOM)*, IEEE, 2019, pp. 24–29.

[54] A. Phosri, K. Ueda, V. L. H. Phung, B. Tawatsupa, A. Honda, and H. Takano, "Effects of ambient air pollution on daily hospital admissions for respiratory and cardiovascular diseases in Bangkok, Thailand," *Sci. Total Environ.*, vol. 651, pp. 1144–1153, Feb. 2019. doi: 10.1016/j.scitotenv.2018.09.183.

[55] K. Ravindra, P. Rattan, S. Mor, and A. N. Aggarwal, "Generalized additive models: Building evidence of air pollution, climate change and human health," *Environ. Int.*, vol. 132, p. 104987, Nov. 2019. doi: 10.1016/j.envint.2019.104987.

[56] D. Parida, A. Behera, J. K. Naik, S. Pattanaik, and R. S. Nanda, "Real-time environment monitoring system using ESP8266 and ThingSpeak on internet of things platform," In *2019 International Conference on Intelligent Computing and Control Systems (ICCS)*, IEEE, 2019, pp. 225–229.

[57] H. P. L. de Medeiros and G. Girao, "An IoT-based air quality monitoring platform," In *2020 IEEE International Smart Cities Conference (ISC2)*, IEEE, Sep. 2020, pp. 1–6. doi: 10.1109/ISC251055.2020.9239070.

[58] A. S. Mihăiţă, L. Dupont, O. Chery, M. Camargo, and C. Cai, "Evaluating air quality by combining stationary, smart mobile pollution monitoring and data-driven modelling," *J. Clean. Prod.*, vol. 221, pp. 398–418, Jun. 2019. doi: 10.1016/j.jclepro.2019.02.179.

[59] G. Kolumban-Antal, V. Lasak, R. Bogdan, and B. Groza, "A secure and portable multi-sensor module for distributed air pollution monitoring," *Sensors*, vol. 20, no. 2, p. 403, Jan. 2020. doi: 10.3390/s20020403.

[60] C. Helbig, M. Ueberham, A. M. Becker, H. Marquart, and U. Schlink, "Wearable sensors for human environmental exposure in urban settings," *Curr. Pollut. Reports*, vol. 7, no. 3, pp. 417–433, 2021.

[61] Y. Song, J. Lin, M. Tang, and S. Dong, "An Internet of energy things based on wireless LPWAN," *Engineering*, vol. 3, no. 4, pp. 460–466, 2017.

[62] A. Simo, S. Dzitac, F. M. Frigura-Iliasa, S. Musuroi, P. Andea, and D. Meianu, "Technical solution for a real-time air quality monitoring system," *Int. J. Comput. Commun. Control*, vol. 15, no. 4, pp. 3891, 2020.

[63] S. Ali, T. Glass, B. Parr, J. Potgieter, and F. Alam, "Low cost sensor with IoT LoRaWAN connectivity and machine learning-based calibration for air pollution monitoring," *IEEE Trans. Instrum. Meas.*, vol. 70, pp. 1–11, 2020.

[64] E. Raghuveera, P. Kanakaraja, K. H. Kishore, C. T. Sriya, and B. S. K. T. Lalith, An IoT enabled air quality monitoring system using LoRa and LPWAN," In *2021 5th International Conference on Computing Methodologies and Communication (ICCMC)*, IEEE, 2021, pp. 453–459.

[65] N. Peladarinos *et al.*, "Early warning systems for COVID-19 infections based on low-cost indoor air-quality sensors and LPWANs," *Sensors*, vol. 21, no. 18, p. 6183, 2021.

[66] N. A. A. Husein, A. H. Abd Rahman, and D. P. Dahnil, "Evaluation of LoRa-based air pollution monitoring system," *Int. J. Adv. Comput. Sci. Appl.*, vol. 10, no. 7, pp. 391–396, 2019.

[67] G. Peruzzi and A. Pozzebon, "A review of energy harvesting techniques for Low Power Wide Area Networks (LPWANs)," *Energies*, vol. 13, no. 13, p. 3433, 2020.

[68] D. Ayala-Ruiz, A. Castillo Atoche, E. Ruiz-Ibarra, E. Osorio de la Rosa, and J. Vázquez Castillo, "A self-powered PMFC-based wireless sensor node for smart city applications," *Wirel. Commun. Mob. Comput.*, vol. 2019, pp. 1–10, 2019.

[69] M. Syafrudin, N. L. Fitriyani, G. Alfian, and J. Rhee, "An affordable fast early warning system for edge computing in assembly line," *Appl. Sci.*, vol. 9, no. 1, p. 84, 2018.

[70] M. Ahmed, R. Mumtaz, S. M. H. Zaidi, M. Hafeez, S. A. R. Zaidi, and M. Ahmad, "Distributed fog computing for Internet of Things (IoT) based ambient data processing and analysis," *Electronics*, vol. 9, no. 11, p. 1756, 2020.

[71] S. U. Amin and M. S. Hossain, "Edge intelligence and Internet of Things in healthcare: A survey," *IEEE Access*, vol. 9, pp. 45–59, 2020.

[72] C. Bellinger, M. S. Mohomed Jabbar, O. Zaïane, and A. Osornio-Vargas, "A systematic review of data mining and machine learning for air pollution epidemiology," *BMC Public Health*, vol. 17, pp. 1–19, 2017.

[73] Y. Bao, Z. Tang, H. Li, and Y. Zhang, "Computer vision and deep learning-based data anomaly detection method for structural health monitoring," *Struct. Heal. Monit.*, vol. 18, no. 2, pp. 401–421, 2019.

[74] J. Wan *et al.*, "Wearable IoT enabled real-time health monitoring system," *EURASIP J. Wirel. Commun. Netw.*, vol. 2018, no. 1, pp. 1–10, 2018.

[75] G. Leenders, G. Callebaut, L. Van der Perre, and L. De Strycker, "Multi-RAT IoT—What's to Gain? An energy-monitoring platform," *arXiv Prepr. arXiv2207.07371*, 2022.

[76] Y. Lalle, L. C. Fourati, M. Fourati, and J. P. Barraca, "A comparative study of LoRaWAN, SigFox, and NB-IoT for smart water grid," In *2019 Global Information Infrastructure and Networking Symposium (GIIS)*, IEEE, Dec. 2019, pp. 1–6. doi: 10.1109/GIIS48668.2019.9044961.

[77] Y. Lalle, M. Fourati, L. C. Fourati, and J. P. Barraca, "Routing strategies for LoRaWAN multi-hop networks: A survey and an SDN-based solution for smart water grid," *IEEE Access*, vol. 9, pp. 168624–168647, 2021.

[78] W. Zhou, Z. Tong, Z. Y. Dong, and Y. Wang, "LoRa-hybrid: A LoRaWAN based multihop solution for regional microgrid," In *2019 IEEE 4th International Conference on Computer and Communication Systems (ICCCS)*, IEEE, Feb. 2019, pp. 650–654. doi: 10.1109/CCOMS.2019.8821683.

[79] M. Pan, C. Chen, X. Yin, and Z. Huang, "UAV-aided emergency environmental monitoring in infrastructure-less areas: LoRa mesh networking approach," *IEEE Internet Things J.*, vol. 9, no. 4, pp. 2918–2932, Feb. 2022. doi: 10.1109/JIOT.2021.3095494.

[80] M. I. Hossain and J. I. Markendahl, "Comparison of LPWAN technologies: Cost structure and scalability," *Wirel. Pers. Commun.*, vol. 121, no. 1, pp. 887–903, 2021.

[81] A. Lavric, A. I. Petrariu, and V. Popa, "Long range SigFox communication protocol scalability analysis under large-scale, high-density conditions," *IEEE Access*, vol. 7, pp. 35816–35825, 2019. doi: 10.1109/ACCESS.2019.2903157.

[82] A. Khalifeh, K. A. Aldahdouh, K. A. Darabkh, and W. Al-Sit, "A survey of 5G emerging wireless technologies featuring LoRaWAN, Sigfox, NB-IoT and LTE-M," In *2019 International Conference on Wireless Communications Signal Processing and Networking (WiSPNET)*, IEEE, Mar. 2019, pp. 561–566. doi: 10.1109/WiSPNET45539.2019.9032817.

[83] L. F. Livi and J. Catani, "A new remote monitor and control system based on SigFox IoT network," *Rev. Sci. Instrum.*, vol. 92, no. 9, p. 094705, Sep. 2021. doi: 10.1063/5.0060336.

[84] S. Ugwuanyi, G. Paul, and J. Irvine, "Survey of IoT for developing countries: Performance analysis of LoRaWAN and cellular NB-IoT networks," *Electronics*, vol. 10, no. 18, p. 2224, Sep. 2021. doi: 10.3390/electronics10182224.

[85] M. B. Hassan, E. S. Ali, R. A. Mokhtar, R. A. Saeed, and B. S. Chaudhari, "NB-IoT: concepts, applications, and deployment challenges," In *LPWAN Technologies for IoT and M2M Applications*, Elsevier, 2020, pp. 119–144. doi: 10.1016/B978-0-12-818880-4.00006-5.

[86] H. Singh, N. Mittal, A. Gupta, Y. Kumar, M. Woźniak, and A. Waheed, "Metamaterial integrated folded dipole antenna with low SAR for 4G, 5G and NB-IoT applications," *Electronics*, vol. 10, no. 21, p. 2612, Oct. 2021. doi: 10.3390/electronics10212612.

[87] H. A. H. Alobaidy, M. J. Singh, R. Nordin, N. F. Abdullah, C. Gze Wei, and M. L. Siang Soon, "Real-world evaluation of power consumption and performance of NB-IoT in Malaysia," *IEEE Internet Things J.*, vol. 9, no. 13, pp. 11614–11632, Jul. 2022. doi: 10.1109/JIOT.2021.3131160.

[88] J. de C. Silva, J. J. P. C. Rodrigues, A. M. Alberti, P. Solic, and A. L. L. Aquino, "LoRaWAN—A low power WAN protocol for Internet of Things: A review and opportunities," In *2017 2nd International Multidisciplinary Conference on Computer and Energy Science (SpliTech)*, 2017, pp. 1–6.

[89] W. A. Jabbar, T. Subramaniam, A. E. Ong, M. I. Shu'Ib, W. Wu, and M. A. de Oliveira, "LoRaWAN-based IoT system implementation for long-range outdoor air quality monitoring," *Internet of Things*, vol. 19, p. 100540, Aug. 2022. doi: 10.1016/j.iot.2022.100540.

[90] A. Lavric and A. I. Petrariu, "LoRaWAN communication protocol: The new era of IoT," In *2018 International Conference on Development and Application Systems (DAS)*, IEEE, 2018, pp. 74–77.

[91] J. Lin, Z. Shen, and C. Miao, "Using blockchain technology to build trust in sharing LoRaWAN IoT," In *Proceedings of the 2nd International Conference on Crowd Science and Engineering*, 2017, pp. 38–43.

[92] G. Kannayeram, M. Madhumitha, S. Mahalakshmi, P. Menaga Devi, K. Monika, and N. B. Prakash, "Smart environmental monitoring using LoRaWAN," In *Proceedings of the 3rd International Conference on Communication, Devices and Computing: ICCDC 2021*, Springer, 2022, pp. 513–520.

[93] J. de Carvalho Silva, J. J. P. C. Rodrigues, A. M. Alberti, P. Solic, and A. L. L. Aquino, "LoRaWAN-A low power WAN protocol for Internet of Things: A review and opportunities," In *2017 2nd International multidisciplinary conference on computer and energy science (SpliTech)*, IEEE, 2017, pp. 1–6.

[94] R. Asif, K. Ghanem, and J. Irvine, "Containerization: For over-the-air programming of field deployed Internet-of-Energy based on cost effective LPWAN," In *2020 First International Conference of Smart Systems and Emerging Technologies (SMARTTECH)*, IEEE, 2020, pp. 65–70.

[95] A. Lavric, A. I. Petrariu, E. Coca, and V. Popa, "LoRaWAN analysis from a high-density Internet of Things perspective," In *2020 International Conference on Development and Application Systems (DAS)*, IEEE, 2020, pp. 94–97.

[96] D. Reeves, "How to create a smart city: Future-proofed cities that foster growth and innovation," *IEEE Electrif. Mag.*, vol. 6, no. 2, pp. 34–41, Jun. 2018. doi: 10.1109/MELE.2018.2816841.

[97] Ç. Civelek, "Development of an IoT based (LoRaWAN) tractor tracking system," *J. Agric. Sci.*, vol. 28, no. 3, p. 21, 2022.

[98] M. A. Ertürk, M. A. Aydın, M. T. Büyükakkaşlar, and H. Evirgen, "A survey on LoRaWAN architecture, protocol and technologies," *Futur. Internet*, vol. 11, no. 10, p. 216, 2019.

[99] H. Zhu et al., "Extreme RSS based indoor localization for LoRaWAN with boundary autocorrelation," *IEEE Trans. Ind. Informatics*, vol. 17, no. 7, pp. 4458–4468, Jul. 2021. doi: 10.1109/TII.2020.2996636.

[100] L. Feltrin, C. Buratti, E. Vinciarelli, R. De Bonis, and R. Verdone, "LoRaWAN: Evaluation of link-and system-level performance," *IEEE Internet Things J.*, vol. 5, no. 3, pp. 2249–2258, 2018.

[101] J. Č. Gambiroža, T. Mastelić, P. Šolić, and M. Čagalj, "Capacity in LoRaWAN networks: Challenges and opportunities," In *2019 4th International Conference on Smart and Sustainable Technologies (SpliTech)*, IEEE, 2019, pp. 1–6.

[102] R. Fujdiak, K. Mikhaylov, J. Pospisil, A. Povalac, and J. Misurec, "Insights into the issue of deploying a private LoRaWAN," *Sensors*, vol. 22, no. 5, p. 2042, 2022.

[103] A. Dongare et al., "Charm: Exploiting geographical diversity through coherent combining in low-power wide-area networks," In *2018 17th ACM/IEEE International Conference on Information Processing in Sensor Networks (IPSN)*, IEEE, 2018, pp. 60–71.

[104] A. Lavric, A. I. Petrariu, and V. Popa, "Sigfox communication protocol: The new era of iot?," In *2019 International Conference on Sensing and Instrumentation in IoT Era (ISSI)*, IEEE, 2019, pp. 1–4.

[105] A. Bucchiarone et al., "Smart construction: remote and adaptable management of construction sites through IoT," *IEEE Internet Things Mag.*, vol. 2, no. 3, pp. 38–45, 2019.

[106] P. Ruckebusch, S. Giannoulis, I. Moerman, J. Hoebeke, and E. De Poorter, "Modelling the energy consumption for over-the-air software updates in LPWAN networks: SigFox, LoRa and IEEE 802.15. 4g," *Internet of Things*, vol. 3, pp. 104–119, 2018.

[107] K. Khujamatov, E. Reypnazarov, D. Khasanov, and N. Akhmedov, "Networking and computing in internet of things and cyber-physical systems," In *2020 IEEE 14th International Conference on Application of Information and Communication Technologies (AICT)*, IEEE, 2020, pp. 1–6.

[108] M. A. M. Almuhaya, W. A. Jabbar, N. Sulaiman, and S. Abdulmalek, "A survey on Lorawan technology: Recent trends, opportunities, simulation tools and future directions," *Electronics*, vol. 11, no. 1, p. 164, 2022.

[109] N. Poursafar, M. E. E. Alahi, and S. Mukhopadhyay, "Long-range wireless technologies for IoT applications: A review," In *2017 Eleventh International Conference on Sensing Technology (ICST)*, IEEE, 2017, pp. 1–6.

[110] A. Overmars and S. Venkatraman, "Towards a secure and scalable iot infrastructure: A pilot deployment for a smart water monitoring system," *Technologies*, vol. 8, no. 4, p. 50, 2020.

[111] G. Soós, D. Ficzere, and P. Varga, "The pursuit of nb-iot transmission rate limitations by real-life network measurements," In *2020 43rd International Conference on Telecommunications and Signal Processing (TSP)*, IEEE, 2020, pp. 430–434.

[112] S. A. Gbadamosi, G. P. Hancke, and A. M. Abu-Mahfouz, "Building upon NB-IoT networks: A roadmap towards 5G new radio networks," *IEEE Access*, vol. 8, pp. 188641–188672, 2020.

[113] J. Shi, L. Jin, J. Li, and Z. Fang, "A smart parking system based on NB-IoT and third-party payment platform," In *2017 17th International Symposium on Communications and Information Technologies (ISCIT)*, IEEE, 2017, pp. 1–5.

[114] K. Mikhaylov et al., "Energy efficiency of multi-radio massive machine-type communication (MR-MMTC): Applications, challenges, and solutions," *IEEE Commun. Mag.*, vol. 57, no. 6, pp. 100–106, 2019.

[115] M. Dangana, S. Ansari, Q. H. Abbasi, S. Hussain, and M. A. Imran, "Suitability of NB-IoT for indoor industrial environment: A survey and insights," *Sensors*, vol. 21, no. 16, p. 5284, 2021.

[116] M. Jouhari, N. Saeed, M.-S. Alouini, and E. M. Amhoud, "A survey on scalable LoRaWAN for massive IoT: Recent advances, potentials, and challenges," *IEEE Commun. Surv. Tutorials*, 2023.

[117] D. Chen, N. Chen, X. Zhang, H. Ma, and Z. Chen, "Next-generation soil moisture sensor web: High-density in situ observation over NB-IoT," *IEEE Internet Things J.*, vol. 8, no. 17, pp. 13367–13383, 2021.

[118] W. Manatarinat, S. Poomrittigul, and P. Tantatsanawong, "Narrowband-internet of things (NB-IoT) system for elderly healthcare services," In *2019 5th International Conference on Engineering, Applied Sciences and Technology (ICEAST)*, IEEE, 2019, pp. 1–4.

[119] H. Zhang, J. Li, B. Wen, Y. Xun, and J. Liu, "Connecting intelligent things in smart hospitals using NB-IoT," *IEEE Internet Things J.*, vol. 5, no. 3, pp. 1550–1560, 2018.

[120] S. Anand and R. Regi, "Remote monitoring of water level in industrial storage tanks using NB-IoT," In *2018 International Conference on Communication Information and Computing Technology (ICCICT)*, IEEE, 2018, pp. 1–4.

[121] A. K. Sultania, F. Mahfoudhi, and J. Famaey, "Real-time demand response using NB-IoT," *IEEE Internet Things J.*, vol. 7, no. 12, pp. 11863–11872, 2020.

[122] N. Naik, "LPWAN technologies for IoT systems: choice between ultra narrow band and spread spectrum," In *2018 IEEE International Systems Engineering Symposium (ISSE)*, IEEE, 2018, pp. 1–8.

[123] J. Hindson, "Development and assessment of an integrated low-cost air quality and traffic sensor network: quantifying traffic-related air pollution in Vancouver, Canada." University of British Columbia, 2022.

[124] L. Zhao, W. Wu, and S. Li, "Design and implementation of an IoT-based indoor air quality detector with multiple communication interfaces," *IEEE Internet Things J.*, vol. 6, no. 6, pp. 9621–9632, 2019.

[125] Y. Wang, Y. Huang, and C. Song, "A new smart sensing system using LoRaWAN for environmental monitoring," In *2019 Computing, Communications and IoT Applications (ComComAp)*, IEEE, 2019, pp. 347–351.

[126] H. Bates, M. Pierce, and A. Benter, "Real-time environmental monitoring for aquaculture using a LoRaWAN-based IoT sensor network," *Sensors*, vol. 21, no. 23, p. 7963, 2021.

[127] J. Park, Y. Oh, H. Byun, and C. Kim, "Low cost fine-grained air quality monitoring system using LoRa WAN," In *2019 International Conference on Information Networking (ICOIN)*, IEEE, 2019, pp. 439–441.

[128] F. Pitu and N. C. Gaitan, "Surveillance of SigFox technology integrated with environmental monitoring," In *2020 International Conference on Development and Application Systems (DAS)*, IEEE, 2020, pp. 69–72.

[129] H. H. R. Sherazi, L. A. Grieco, M. A. Imran, and G. Boggia, "Energy-efficient LoRaWAN for industry 4.0 applications," *IEEE Trans. Ind. Informatics*, vol. 17, no. 2, pp. 891–902, 2020.

[130] S. M. Mousavi, A. Khademzadeh, and A. M. Rahmani, "The role of low-power wide-area network technologies in Internet of Things: A systematic and comprehensive review," *Int. J. Commun. Syst.*, vol. 35, no. 3, p. e5036, 2022.

[131] G. G. L. Ribeiro, L. F. de Lima, L. Oliveira, J. J. P. C. Rodrigues, C. N. M. Marins, and G. A. B. Marcondes, "An outdoor localization system based on SigFox," In *2018 IEEE 87th Vehicular Technology Conference (VTC Spring)*, IEEE, 2018, pp. 1–5.

[132] S. J. Johnston et al., "City scale particulate matter monitoring using LoRaWAN based air quality IoT devices," *Sensors*, vol. 19, no. 1, p. 209, 2019.

[133] M. S. H. Sassi and L. C. Fourati, "Comprehensive survey on air quality monitoring systems based on emerging computing and communication technologies," *Comput. Networks*, p. 108904, 2022.

[134] M. Meli, E. Gatt, O. Casha, I. Grech, and J. Micallef, "A novel low power and low cost IoT wireless sensor node for air quality monitoring," In *2020 27th IEEE International Conference on Electronics, Circuits and Systems (ICECS)*, IEEE, 2020, pp. 1–4.

[135] J.-M. Martinez-Caro and M.-D. Cano, "IoT system integrating unmanned aerial vehicles and LoRa technology: A performance evaluation study," *Wirel. Commun. Mob. Comput.*, vol. 2019, pp. 1–12, 2019.

[136] Q. M. Qadir, T. A. Rashid, N. K. Al-Salihi, B. Ismael, A. A. Kist, and Z. Zhang, "Low power wide area networks: A survey of enabling technologies, applications and interoperability needs," *IEEE Access*, vol. 6, pp. 77454–77473, 2018.

[137] B. Rana, Y. Singh, and P. K. Singh, "A systematic survey on internet of things: Energy efficiency and interoperability perspective," *Trans. Emerg. Telecommun. Technol.*, vol. 32, no. 8, p. e4166, 2021.

[138] J. Shi, X. Chen, and M. Sha, "Enabling direct messaging from LoRa to ZigBee in the 2.4 GHz band for industrial wireless networks," In *2019 IEEE International Conference on Industrial Internet (ICII)*, IEEE, 2019, pp. 180–189.

[139] B. Al Homssi *et al.*, "A framework for the design and deployment of large-scale LPWAN networks for smart cities applications," *IEEE Internet Things Mag.*, vol. 4, no. 1, pp. 53–59, 2020.

[140] W. Xu *et al.*, "The design, implementation, and deployment of a smart lighting system for smart buildings," *IEEE Internet Things J.*, vol. 6, no. 4, pp. 7266–7281, 2019.

[141] B. Thoen, G. Callebaut, G. Leenders, and S. Wielandt, "A deployable LPWAN platform for low-cost and energy-constrained IoT applications," *Sensors*, vol. 19, no. 3, p. 585, 2019.

[142] R. Sanchez-Iborra, "LPWAN and embedded machine learning as enablers for the next generation of wearable devices," *Sensors*, vol. 21, no. 15, p. 5218, 2021.

[143] C. A. Hernández-Morales, J. M. Luna-Rivera, and R. Perez-Jimenez, "Design and deployment of a practical IoT-based monitoring system for protected cultivations," *Comput. Commun.*, vol. 186, pp. 51–64, 2022.

[144] S. Idris, T. Karunathilake, and A. Förster, "Survey and comparative study of LoRa-enabled simulators for internet of things and wireless sensor networks," *Sensors*, vol. 22, no. 15, p. 5546, 2022.

[145] G. Alsuhli, A. Fahim, and Y. Gadallah, "A survey on the role of UAVs in the communication process: A technological perspective," *Comput. Commun.*, vol. 194, pp. 86–123, 2022.

[146] S. Duangsuwan, A. Takarn, and P. Jamjareegulgarn, "A development on air pollution detection sensors based on NB-IoT network for smart cities," In *2018 18th International Symposium on Communications and Information Technologies (ISCIT)*, IEEE, 2018, pp. 313–317.

[147] S. R. Niya, S. S. Jha, T. Bocek, and B. Stiller, "Design and implementation of an automated and decentralized pollution monitoring system with blockchains, smart contracts, and LoRaWAN," In *NOMS 2018-2018 IEEE/IFIP Network Operations and Management Symposium*, IEEE, 2018, pp. 1–4.

[148] K.-D. Walter, "AI-based sensor platforms for the IoT in smart cities," In *Big Data Analytics for Cyber-Physical Systems*, Elsevier, 2019, pp. 145–166.

[149] T. Zhuang, M. Ren, X. Gao, M. Dong, W. Huang, and C. Zhang, "Insulation condition monitoring in distribution power grid via IoT-based sensing network," *IEEE Trans. Power Deliv.*, vol. 34, no. 4, pp. 1706–1714, 2019.

[150] J. Jose and T. Sasipraba, "Indoor air quality monitors using IOT sensors and LPWAN," In *2019 3rd International conference on Trends in electronics and informatics (ICOEI)*, IEEE, 2019, pp. 633–637.

[151] M. Saravanan, A. Das, and V. Iyer, "Smart water grid management using LPWAN IoT technology," In *2017 Global Internet of Things Summit (GIoTS)*, IEEE, 2017, pp. 1–6.

[152] D. Loukatos, I. Manolopoulos, E.-S. Arvaniti, K. G. Arvanitis, and N. A. Sigrimis, "Experimental testbed for monitoring the energy requirements of LPWAN equipped sensor nodes," *IFAC-PapersOnLine*, vol. 51, no. 17, pp. 309–313, 2018.

[153] R. K. Singh, P. P. Puluckul, R. Berkvens, and M. Weyn, "Energy consumption analysis of LPWAN technologies and lifetime estimation for IoT application," *Sensors*, vol. 20, no. 17, p. 4794, 2020.

[154] D.-Y. Kim, S. Kim, H. Hassan, and J. H. Park, "Radio resource management for data transmission in low power wide area networks integrated with large scale cyber physical systems," *Cluster Comput.*, vol. 20, pp. 1831–1842, 2017.

[155] F. Wu, T. Wu, and M. R. Yuce, "An internet-of-things (IoT) network system for connected safety and health monitoring applications," *Sensors*, vol. 19, no. 1, p. 21, 2018.

[156] A. Matese, S. F. Di Gennaro, A. Zaldei, L. Genesio, and F. P. Vaccari, "A wireless sensor network for precision viticulture: The NAV system," *Comput. Electron. Agric.*, vol. 69, no. 1, pp. 51–58, 2009.

[157] L. Schor, P. Sommer, and R. Wattenhofer, "Towards a zero-configuration wireless sensor network architecture for smart buildings," In *Proceedings of the First ACM Workshop on Embedded Sensing Systems for Energy-Efficiency in Buildings*, 2009, pp. 31–36.

[158] A. Augustin, J. Yi, T. Clausen, and W. M. Townsley, "A study of LoRa: Long range & low power networks for the internet of things," *Sensors*, vol. 16, no. 9, p. 1466, 2016.

[159] W.-J. Yi, W. Jia, and J. Saniie, "Mobile sensor data collector using Android smartphone," In *2012 IEEE 55th International Midwest Symposium on Circuits and Systems (MWSCAS)*, IEEE, 2012, pp. 956–959.

[160] M. Li, W. Lou, and K. Ren, "Data security and privacy in wireless body area networks," *IEEE Wirel. Commun.*, vol. 17, no. 1, pp. 51–58, 2010.

[161] B.-R. Chen et al., "A web-based system for home monitoring of patients with Parkinson's disease using wearable sensors," *IEEE Trans. Biomed. Eng.*, vol. 58, no. 3, pp. 831–836, 2010.

[162] P. Chen, "Visualization of real-time monitoring datagraphic of urban environmental quality," *Eurasip J. Image Video Process.*, vol. 2019, no. 1, pp. 1–9, 2019.

[163] B. Braem, S. Latré, P. Leroux, P. Demeester, T. Coenen, and P. Ballon, "Designing a smart city playground: Real-time air quality measurements and visualization in the City of Things testbed," In *2016 IEEE International Smart Cities Conference (ISC2)*, IEEE, 2016, pp. 1–2.

[164] J. C. K. Lam, L. Y. L. Cheung, S. Wang, and V. O. K. Li, "Stakeholder concerns of air pollution in Hong Kong and policy implications: A big-data computational text analysis approach," *Environ. Sci. Policy*, vol. 101, pp. 374–382, 2019.

[165] T. Montanaro, I. Sergi, M. Basile, L. Mainetti, and L. Patrono, "An IOT-aware solution to support governments in air pollution monitoring based on the combination of real-time data and citizen feedback," *Sensors*, vol. 22, no. 3, p. 1000, 2022.

[166] O. H. Kombo, S. Kumaran, and A. Bovim, "Design and application of a low-cost, low-power, LoRa-GSM, IoT enabled system for monitoring of groundwater resources with energy harvesting integration," *IEEE Access*, vol. 9, pp. 128417–128433, 2021.

[167] M. Reyneke, B. Mullins, and M. Reith, "LoRaWAN & The Helium block-chain: A study on military IoT deployment," In *International Conference on Cyber Warfare and Security*, 2023, pp. 327–337.

[168] S. Liu, C. Xia, and Z. Zhao, "A low-power real-time air quality monitoring system using LPWAN based on LoRa," In *2016 13th IEEE International Conference on Solid-State and Integrated Circuit Technology (ICSICT)*, IEEE, 2016, pp. 379–381.

[169] N. Koyana, E. D. Markus, and A. M. Abu-Mahfouz, "A survey of network and intelligent air pollution monitoring in South Africa," In *2019 International Multidisciplinary Information Technology and Engineering Conference (IMITEC)*, IEEE, 2019, pp. 1–7.

[170] B. Buurman, J. Kamruzzaman, G. Karmakar, and S. Islam, "Low-power wide-area networks: Design goals, architecture, suitability to use cases and research challenges," *IEEE Access*, vol. 8, pp. 17179–17220, 2020.

[171] S. Le et al., "Use of low-cost air quality sensors and satellite observations in Uganda for improved air pollution exposure assessment," In *AGU Fall Meeting Abstracts*, 2022, pp. A42O–1912.

[172] W. Ayoub, A. E. Samhat, F. Nouvel, M. Mroue, H. Jradi, and J.-C. Prévotet, "Media independent solution for mobility management in heterogeneous LPWAN technologies," *Comput. Networks*, vol. 182, p. 107423, 2020.

[173] K. F. Muteba, K. Djouani, and T. Olwal, "5G NB-IoT: Design, considerations, solutions and challenges," *Procedia Comput. Sci.*, vol. 198, pp. 86–93, 2022.

[174] S. R. Borkar, "Long-term evolution for machines (LTE-M)," In *LPWAN Technologies for IoT and M2M Applications*, Elsevier, 2020, pp. 145–166.

[175] D.-Y. Kim, S. Kim, H. Hassan, and J. H. Park, "Adaptive data rate control in low power wide area networks for long range IoT services," *J. Comput. Sci.*, vol. 22, pp. 171–178, 2017.

[176] D. Trnka, "Policies, regulatory framework and enforcement for air quality management: The case of Korea," 2020.

[177] Y. Kabalci, E. Kabalci, S. Padmanaban, J. B. Holm-Nielsen, and F. Blaabjerg, "Internet of things applications as energy internet in smart grids and smart environments," *Electronics*, vol. 8, no. 9, p. 972, 2019.

[178] B. S. Chaudhari and M. Zennaro, *LPWAN technologies for IoT and M2M applications*. Academic Press, 2020.

[179] B. Chaparro, F.; Pérez, M.; Mendez, D. "A communication framework for image transmission through LPWAN technology." *Electronics*, 2022, 11, 1764. s Note: MDPI stays neutral with regard to jurisdictional claims in pub-lished ..., 2022.

Chapter 8

Review of AI techniques for LPWAN sensors
Enhancing efficiency and intelligence

Zakaria Boulouard
LIM, Hassan II University

Mariya Ouaissa
Cadi Ayyad University

Mariyam Ouaissa
Chouaib Doukkali University

Sarah El Himer
Sidi Mohammed Ben Abdellah University

8.1 INTRODUCTION

The Internet of Things (IoT) has witnessed remarkable growth, enabling the deployment of various sensors for monitoring environmental parameters. Among the IoT technologies, Low Power Wide Area Network (LPWAN) has emerged as a popular choice due to its cost-effectiveness and suitability for resource-constrained sensor nodes [1]. LPWAN sensors provide a scalable and efficient solution for collecting data from remote locations over long distances while operating on low power [2].

However, the widespread adoption of LPWAN sensors has given rise to a new challenge: the overwhelming volume of data generated by these sensors [3]. Analyzing and extracting meaningful insights from such vast amounts of data has become increasingly complex. Fortunately, Artificial Intelligence (AI) techniques have emerged as a powerful toolset to address these challenges and unlock the full potential of LPWAN sensor data [4].

This review paper aims to provide a comprehensive overview of the state-of-the-art AI techniques applied in the context of LPWAN sensors. By leveraging AI techniques, LPWAN sensors can not only efficiently collect data but also analyze it in real-time, enabling intelligent decision-making and proactive resource management [5–7]. Through this review, we explore the various AI algorithms and models that have been employed for

DOI: 10.1201/9781003426974-8

data analysis in LPWAN networks and investigate their applications and benefits.

The objectives of this review are threefold: first, to present an overview of LPWAN technology, highlighting its characteristics and advantages for IoT deployments; second, to discuss the challenges associated with LPWAN sensor data, including data volume, noise, and energy constraints; and third, to explore the diverse range of AI techniques, such as machine learning algorithms and deep learning models, that have been applied to LPWAN sensor data analysis. In addition, we will examine specific applications of AI in LPWAN sensor networks, showcasing how it enhances efficiency, enables predictive maintenance, optimizes energy consumption, and detects anomalies.

By providing insights into the application of AI techniques in LPWAN sensors, this review paper aims to contribute to the existing body of knowledge and provide valuable guidance to researchers, practitioners, and policymakers working on LPWAN-based IoT systems. Furthermore, we will highlight the future research directions and address the challenges and opportunities in adopting AI for LPWAN sensor networks.

Overall, this review paper serves as a foundation for understanding the potential of AI techniques in enhancing the efficiency and intelligence of LPWAN sensors, leading to more effective and insightful utilization of data in LPWAN deployments.

8.2 OVERVIEW OF LPWAN TECHNOLOGY

8.2.1 Definition and characteristics

LPWAN technology is a wireless communication protocol specifically designed to enable long-range connectivity for IoT devices [8]. LPWAN offers distinct advantages in terms of cost-effectiveness, long battery life, and extended coverage. It operates on an unlicensed radio spectrum, allowing for easy deployment and scalability.

8.2.2 Types of LPWAN

There are several LPWAN technologies available, each with its own characteristics and use cases. The prominent LPWAN technologies include LoRaWAN, Sigfox, NB-IoT (Narrowband IoT), and LTE-M (Long-Term Evolution for Machines). LoRaWAN, based on the LoRa modulation technique, provides low-power, long-range communication with support for massive device connectivity. Sigfox, on the other hand, utilizes ultra-narrowband technology to offer low-cost, low-power, and long-range connectivity. NB-IoT and LTE-M are cellular-based LPWAN technologies, leveraging existing cellular infrastructure to provide wide coverage, deep penetration, and higher data rates [9].

8.2.3 Advantages of LPWAN sensors

LPWAN sensors have gained significant popularity in various IoT applications due to their unique advantages. First, LPWAN sensors require minimal power, allowing for long-lasting battery life, which is crucial for remote and battery-powered deployments [10]. Second, LPWAN offers extended coverage, enabling communication over several kilometers, making it suitable for wide-scale monitoring and tracking applications. In addition, LPWAN sensors provide cost-effective solutions by minimizing infrastructure requirements and operating on unlicensed frequency bands [11].

8.2.4 Use cases of LPWAN sensors

LPWAN sensors find application in diverse fields, including agriculture, environmental monitoring, smart cities, asset tracking, and industrial automation. In agriculture, LPWAN sensors enable farmers to monitor soil moisture, temperature, and other parameters, facilitating precise irrigation and optimizing crop yield [12]. Environmental monitoring applications utilize LPWAN sensors for tracking air quality [13], water quality [14], and weather conditions [15]. LPWAN sensors are also employed in asset tracking systems to monitor the location and condition of valuable assets in real-time [16].

8.2.5 LPWAN ecosystem and standards

The LPWAN ecosystem consists of sensor nodes, gateways, network servers, and application servers. Sensor nodes collect data from the environment, which is then transmitted to gateways. Gateways act as intermediaries, relaying the data to network servers where it is processed and forwarded to application servers. LPWAN standards and protocols ensure interoperability and compatibility among different LPWAN networks, allowing seamless integration of devices and systems [9].

Understanding the fundamentals of LPWAN technology and its advantages is crucial for realizing the potential of LPWAN sensors in IoT applications. In the following sections, we will delve into the challenges associated with LPWAN sensor data and explore how AI techniques can enhance the efficiency and intelligence of LPWAN sensor networks.

8.3 CHALLENGES IN LPWAN SENSOR DATA

8.3.1 Data volume

LPWAN sensor networks generate a massive amount of data due to their wide coverage and continuous monitoring capabilities. This high data volume poses challenges in terms of storage, transmission, and processing [3]. Traditional data analysis techniques may not be suitable for handling such large-scale data sets efficiently. Therefore, innovative approaches are required to address the data volume challenge and extract meaningful insights from the collected data [17].

8.3.2 Noise and data quality

LPWAN sensors operate in diverse environmental conditions, which can introduce various sources of noise into the collected data. Factors such as interference, signal attenuation, and environmental disturbances can impact the quality and reliability of the sensor data. The presence of noise can hinder accurate analysis and interpretation, leading to potential errors in decision-making. Addressing noise and ensuring data quality are essential for obtaining reliable results from LPWAN sensor networks [18,19].

8.3.3 Energy constraints

LPWAN sensors are typically battery-powered and designed to operate for extended periods without frequent maintenance. Energy efficiency is critical to prolong the lifespan of sensor nodes and reduce the need for frequent battery replacements. However, energy constraints pose challenges in terms of data collection, transmission, and processing [20]. Energy-efficient algorithms and techniques need to be developed to optimize energy consumption while maintaining the desired level of data accuracy and analysis [21,22].

Addressing these challenges associated with LPWAN sensor data is vital to fully leverage the benefits offered by LPWAN technology. In the following sections, we explore how AI techniques can be employed to overcome these challenges and enhance the efficiency and intelligence of LPWAN sensor networks.

8.4 AI TECHNIQUES FOR LPWAN SENSOR DATA ANALYSIS

8.4.1 Machine learning algorithms

Machine learning algorithms have gained prominence in analyzing LPWAN sensor data due to their ability to discover patterns, make predictions, and classify data. Supervised learning algorithms, such as support vector machines (SVM) and Random Forests, can be utilized for regression tasks [23,24]. Unsupervised learning algorithms, like k-means clustering and self-organizing maps (SOM), can uncover hidden patterns and structures within the data [25,26]. Reinforcement learning algorithms, such as

Q-Learning and Deep Q-Networks (DQN) [19,27], enable LPWAN sensors to learn optimal decision-making policies through interaction with the environment.

8.4.2 Deep learning models

Deep learning models, specifically neural networks, have shown remarkable capabilities in processing and analyzing complex data. Convolutional neural networks (CNNs) can extract spatial and temporal features from LPWAN sensor data, making them suitable for tasks such as image recognition and signal processing. Recurrent neural networks (RNNs) and long short-term memory (LSTM) networks are effective in modeling sequential and time-series data, allowing LPWAN sensors to capture dependencies and make predictions based on historical sensor readings [28,29].

8.4.3 Data fusion and multi-sensor integration

LPWAN sensor networks often comprise multiple sensors deployed in different locations. AI techniques enable the fusion of data from multiple sensors to provide a holistic view of the monitored environment. Data fusion algorithms, such as Kalman filters and Particle filters, combine sensor measurements to estimate the true state of the system accurately [30]. Multi-sensor integration techniques, leveraging AI models, can enhance the accuracy and reliability of data analysis by integrating information from diverse sensor sources [31].

8.4.4 Edge computing and distributed AI

LPWAN sensor networks can benefit from edge computing and distributed AI techniques to overcome latency and bandwidth limitations. Edge computing involves performing data analysis and AI computations at the edge of the network, closer to the sensor nodes, reducing the need for data transmission to centralized servers. This approach enables real-time decision-making and reduces the burden on the LPWAN network [32]. Additionally, distributed AI techniques, such as federated learning and collaborative learning, facilitate collaborative model training and knowledge sharing among LPWAN sensor nodes while preserving data privacy and security [33].

By leveraging these AI techniques, LPWAN sensors can efficiently analyze data, detect anomalies, make predictions, and optimize resource utilization. The next section explores diverse applications of AI in LPWAN sensor networks, highlighting how these techniques enhance efficiency and intelligence in various IoT domains.

8.5 APPLICATIONS OF AI IN LPWAN SENSOR NETWORKS

8.5.1 Anomaly detection

AI techniques play a crucial role in identifying anomalies and deviations from normal behavior in LPWAN sensor data. By learning patterns and trends from historical data, AI models can detect unusual events or outliers that may indicate equipment malfunctions, environmental hazards, or security breaches. Anomaly detection enables proactive maintenance, early warning systems, and improved operational efficiency in various domains, such as industrial monitoring, infrastructure management, and security surveillance [34].

8.5.2 Predictive maintenance

LPWAN sensors coupled with AI algorithms enable predictive maintenance, a proactive approach to asset management. By analyzing sensor data and identifying patterns that precede equipment failures or performance degradation, AI models can predict when maintenance is required. This empowers organizations to schedule maintenance activities, minimize downtime, optimize resource allocation, and reduce maintenance costs [35].

8.5.3 Data compression and transmission

LPWAN networks often face bandwidth constraints due to low data rates. AI techniques, such as data compression and feature extraction, enable efficient data transmission and storage. By reducing the dimensionality of sensor data or using compression algorithms like wavelet transforms, AI can significantly reduce the amount of data transmitted over the LPWAN network, optimizing bandwidth utilization without sacrificing critical information [36,37].

8.5.4 Energy optimization

LPWAN sensors operate on limited battery power, making energy optimization a crucial aspect. AI techniques can assist in energy-efficient data collection, transmission, and processing. For example, machine learning models can be trained to make intelligent decisions about when to activate or deactivate sensors based on environmental conditions, minimizing energy consumption while maintaining data accuracy. AI can also optimize routing protocols and resource allocation, maximizing the lifespan of sensor nodes [38,39].

8.5.5 Environmental monitoring and control

LPWAN sensors combined with AI capabilities enable intelligent environmental monitoring and control systems. By analyzing data from various sensors measuring temperature and humidity [40], air and water quality [6,13], as well as other parameters, AI models can provide insights into environmental conditions. This information can be utilized to optimize energy consumption, regulate indoor climate, manage greenhouse environments [41], and support sustainable practices in agriculture [12,42] and urban planning [1].

AI-driven applications in LPWAN sensor networks continue to expand, offering solutions that enhance operational efficiency, improve decision-making, and enable intelligent automation across diverse domains. The following section discusses future directions and research opportunities in the field of AI for LPWAN sensors.

8.6 FUTURE DIRECTIONS AND RESEARCH OPPORTUNITIES

8.6.1 Integration with edge intelligence

The integration of AI techniques with edge intelligence holds promising potential for LPWAN sensor networks. By deploying AI models directly on edge devices or gateways, real-time data analysis and decision-making can be achieved without relying heavily on cloud infrastructure. Future research can focus on developing lightweight AI models that can operate efficiently on resource-constrained edge devices, enabling faster response times and reducing reliance on centralized processing.

8.6.2 Federated learning for privacy-preserving collaboration

Privacy is a critical concern in LPWAN sensor networks, as they often collect sensitive data. Federated learning, a distributed AI approach, enables collaborative model training across multiple LPWAN sensor nodes while preserving data privacy. Future research can explore advanced techniques for federated learning in LPWAN networks, addressing challenges such as communication overhead, model aggregation, and security, to enhance privacy and collaboration among distributed LPWAN sensors.

8.6.3 Hybrid AI approaches

Hybrid AI approaches that combine multiple AI techniques can unlock new opportunities for LPWAN sensor data analysis. For example, integrating machine learning with symbolic reasoning or knowledge representation techniques can enhance the interpretability and explainability of AI models. Hybrid models that combine deep learning with rule-based systems can leverage the strengths of both approaches, enabling accurate analysis and decision-making while providing human-understandable explanations.

8.6.4 Adaptive and self-learning systems

LPWAN sensor networks operate in dynamic environments where conditions and requirements may change over time. Developing adaptive and self-learning systems for LPWAN sensors can enable them to autonomously adapt to changing circumstances, optimize performance, and improve accuracy. Future research can explore techniques such as online learning, reinforcement learning, and transfer learning to develop intelligent LPWAN sensor systems that continually evolve and adapt to new scenarios.

8.6.5 Integration with blockchain technology

The integration of LPWAN sensor networks with blockchain technology offers opportunities for secure and transparent data management, integrity verification, and decentralized governance. Blockchain can enhance trust, data immutability, and auditability in LPWAN sensor networks, ensuring data integrity and enabling secure data sharing and monetization. Future research can focus on exploring the potential of blockchain-based architectures and smart contracts in LPWAN sensor networks.

As AI continues to advance, there are numerous research directions and opportunities to explore in the field of LPWAN sensor networks. The development of innovative techniques, algorithms, and frameworks can further enhance the efficiency, intelligence, and scalability of LPWAN-based IoT systems, ultimately enabling transformative applications and advancements in various domains.

8.7 CONCLUSION

LPWAN sensor networks, with their low power consumption, wide coverage, and long-range connectivity, have emerged as a prominent technology for enabling large-scale IoT deployments. However, the analysis and utilization of the vast amount of data generated by LPWAN sensors present unique challenges. In this paper, we explored the potential of AI techniques in addressing these challenges and enhancing the efficiency and intelligence of LPWAN sensor networks.

We discussed how machine learning algorithms and deep learning models can be applied to analyze LPWAN sensor data, enabling tasks such as anomaly detection, predictive maintenance, and data compression. We also highlighted the importance of data fusion and multi-sensor integration in achieving a comprehensive understanding of the monitored environment. Furthermore, we explored the benefits of edge computing, distributed AI, and the integration of AI with LPWAN technology.

The applications of AI in LPWAN sensor networks are diverse and impactful. From anomaly detection and predictive maintenance to energy optimization and environmental monitoring, AI-driven solutions enable efficient operations, informed decision-making, and resource optimization across various domains. Moreover, we discussed future research directions and opportunities, including the integration of edge intelligence, federated learning, hybrid AI approaches, adaptive systems, and blockchain technology.

As LPWAN technology continues to advance and AI techniques evolve, there is a need for continued research and innovation to fully unlock the potential of LPWAN sensor networks. Further exploration of advanced AI algorithms, efficient edge computing frameworks, and privacy-preserving collaboration techniques can contribute to the development of intelligent, scalable, and secure LPWAN-based IoT systems.

In conclusion, AI techniques have proven to be valuable tools in addressing the challenges of LPWAN sensor data analysis and unlocking new opportunities for smart applications. The combination of LPWAN technology and AI capabilities has the potential to revolutionize various industries, improve resource utilization, and enable data-driven decision-making. It is an exciting and evolving field, and further research and collaboration are essential to harness the full potential of AI in LPWAN sensor networks and drive the next wave of IoT innovation.

REFERENCES

[1] E. U. Ogbodo, A. M. Abu-Mahfouz, and A. M. Kurien, "A Survey on 5G and LPWAN-IoT for Improved Smart Cities and Remote Area Applications: From the Aspect of Architecture and Security," *Sensors*, vol. 22, no. 16, p. 6313, Aug. 2022. doi: 10.3390/s22166313.

[2] B. S. Chaudhari and M. Zennaro, *LPWAN Technologies for IoT and M2M Applications*. Elsevier, 2020. doi: 10.1016/C2018-0-04787-8.

[3] Y. Chen, Y. A. Sambo, O. Onireti, and M. A. Imran, "A Survey on LPWAN-5G Integration: Main Challenges and Potential Solutions," *IEEE Access*, vol. 10, pp. 32132–32149, 2022. doi: 10.1109/ACCESS.2022.3160193.

[4] Z. Boulouard, M. Ouaissa, M. Ouaissa, F. Siddiqui, M. Almutiq, and M. Krichen, "An Integrated Artificial Intelligence of Things Environment for River Flood Prevention," *Sensors*, vol. 22, no. 23, Dec. 2022. doi: 10.3390/s22239485.

[5] P. Singh, Z. Elmi, V. Krishna Meriga, J. Pasha, and M. A. Dulebenets, "Internet of Things for Sustainable Railway Transportation: Past, Present, and Future," *Clean. Logist. Supply Chain*, vol. 4, no. June, p. 100065, Jul. 2022. doi: 10.1016/j.clscn.2022.100065.

[6] P. Boccadoro, V. Daniele, P. Di Gennaro, D. Lofù, and P. Tedeschi, "Water Quality Prediction on a Sigfox-Compliant IoT Device: The Road Ahead of Waters" *Ad Hoc Networks*, vol. 126, no. November 2021, p. 102749, Mar. 2022. doi: 10.1016/j.adhoc.2021.102749.

[7] G. Signoretti, M. Silva, P. Andrade, I. Silva, E. Sisinni, and P. Ferrari, "An Evolving TinyML Compression Algorithm for IoT Environments Based on Data Eccentricity," *Sensors*, vol. 21, no. 12, p. 4153, Jun. 2021. doi: 10.3390/s21124153.

[8] C. Milarokostas, D. Tsolkas, N. Passas, and L. Merakos, "A Comprehensive Study on LPWANs with a Focus on the Potential of LoRa/LoRaWAN Systems," *IEEE Commun. Surv. Tutorials*, vol. 25, no. 1, pp. 825–867, 2023. doi: 10.1109/COMST.2022.3229846.

[9] B. S. Chaudhari, M. Zennaro, and S. Borkar, "LPWAN Technologies: Emerging Application Characteristics, Requirements, and Design Considerations," *Futur. Internet*, vol. 12, no. 3, p. 46, Mar. 2020. doi: 10.3390/fi12030046.

[10] R. K. Singh, P. P. Puluckul, R. Berkvens, and M. Weyn, "Energy Consumption Analysis of LPWAN Technologies and Lifetime Estimation for IoT Application," *Sensors*, vol. 20, no. 17, p. 4794, Aug. 2020. doi: 10.3390/s20174794.

[11] S. Idris, T. Karunathilake, and A. Förster, "Survey and Comparative Study of LoRa-Enabled Simulators for Internet of Things and Wireless Sensor Networks," *Sensors*, vol. 22, no. 15, p. 5546, Jul. 2022. doi: 10.3390/s22155546.

[12] K. Paul et al., "Viable Smart Sensors and Their Application in Data Driven Agriculture," *Comput. Electron. Agric.*, vol. 198, no. May, p. 107096, Jul. 2022. doi: 10.1016/j.compag.2022.107096.

[13] E. Raghuveera, P. Kanakaraja, K. H. Kishore, C. Tanvi Sriya, D. P. B, and B. Sai Krishna Teja Lalith, "An IoT Enabled Air Quality Monitoring System Using LoRa and LPWAN," In *2021 5th International Conference on Computing Methodologies and Communication (ICCMC)*, Apr. 2021, pp. 453–459. doi: 10.1109/ICCMC51019.2021.9418440.

[14] E. C. Okpara, B. E. Sehularo, and O. B. Wojuola, "On-line Water Quality Inspection System: The Role of the Wireless Sensory Network," *Environ. Res. Commun.*, vol. 4, no. 10, p. 102001, Oct. 2022. doi: 10.1088/2515-7620/ac9aa5.

[15] A. Udomlumlert, N. Vichaidis, P. Kamin, T. Surasak, S. C. H. Huang, and N. Jiteurtragool, "The Development of Air Quality Monitoring System using IoT and LPWAN," In *2020-5th International Conference on Information Technology (InCIT)*, Oct. 2020, pp. 254–258. doi: 10.1109/InCIT50588.2020.9310949.

[16] T. Janssen, A. Koppert, R. Berkvens, and M. Weyn, "A Survey on IoT Positioning leveraging LPWAN, GNSS and LEO-PNT," *IEEE Internet Things J.*, vol. PP, no. X, pp. 1–1, 2023. doi: 10.1109/JIOT.2023.3243207.

[17] R. Kufakunesu, G. P. Hancke, and A. M. Abu-Mahfouz, "A Survey on Adaptive Data Rate Optimization in LoRaWAN: Recent Solutions and Major Challenges," *Sensors*, vol. 20, no. 18, p. 5044, Sep. 2020. doi: 10.3390/s20185044.

[18] A. P. Matz, J.-A. Fernandez-Prieto, J. Canada-Bago, and U. Birkel, "The Narrowband Bundling Protocol," *IEEE Wirel. Commun. Lett.*, vol. 11, no. 9, pp. 1900–1904, Sep. 2022. doi: 10.1109/LWC.2022.3186392.

[19] O. J. Pandey, T. Yuvaraj, J. K. Paul, H. H. Nguyen, K. Gundepudi, and M. K. Shukla, "Improving Energy Efficiency and QoS of LPWANs for IoT Using Q-Learning Based Data Routing," *IEEE Trans. Cogn. Commun. Netw.*, vol. 8, no. 1, pp. 365–379, Mar. 2022. doi: 10.1109/TCCN.2021.3114147.

[20] T. Perković, S. Damjanović, P. Šolić, L. Patrono, and J. J. P. C. Rodrigues, "Meeting Challenges in IoT: Sensing, Energy Efficiency, and the Implementation," Adv. Intell. Syst. Comput., vol. 1041, 2020, pp. 419–430. doi: 10.1007/978-981-15-0637-6_36.

[21] Z. E. Ahmed, R. A. Saeed, A. Mukherjee, and S. N. Ghorpade, "Energy Optimization in Low-Power Wide Area Networks by Using Heuristic Techniques," In *LPWAN Technologies for IoT and M2M Applications*, Elsevier, 2020, pp. 199–223. doi: 10.1016/B978-0-12-818880-4.00011-9.

[22] H. Wang, Y. Wu, Y. Liu, and W. Liu, "A Cross-Layer Framework for LPWAN Management based on Fuzzy Cognitive Maps with Adaptive Glowworm Swarm Optimization," In *2023 IEEE Wireless Communications and Networking Conference (WCNC)*, Mar. 2023, pp. 1–6. doi: 10.1109/WCNC55385.2023.10119004.

[23] H. Rajab, M. Ebrahim, and T. Cinkler, "Reducing Power Requirement of LPWA Networks via Machine Learning," *Pollack Period.*, vol. 16, no. 2, pp. 86–91, Jun. 2021. doi: 10.1556/606.2020.00263.

[24] M. Anjum, M. A. Khan, S. A. Hassan, A. Mahmood, H. K. Qureshi, and M. Gidlund, "RSSI Fingerprinting-Based Localization Using Machine Learning in LoRa Networks," *IEEE Internet Things Mag.*, vol. 3, no. 4, pp. 53–59, Dec. 2020. doi: 10.1109/IOTM.0001.2000019.

[25] M. S. A. Muthanna, P. Wang, M. Wei, A. Rafiq, and N. N. Josbert, "Clustering Optimization of LoRa Networks for Perturbed Ultra-Dense IoT Networks," *Information*, vol. 12, no. 2, p. 76, Feb. 2021. doi: 10.3390/info12020076.

[26] M. O. Farooq, "Clustering-Based Layering Approach for Uplink Multi-Hop Communication in LoRa Networks," *IEEE Netw. Lett.*, vol. 2, no. 3, pp. 132–135, Sep. 2020. doi: 10.1109/LNET.2020.3003161.

[27] A. Kaburaki, K. Adachi, O. Takyu, M. Ohta, and T. Fujii, "Autonomous Decentralized Traffic Control Using Q-Learning in LPWAN," *IEEE Access*, vol. 9, pp. 93651–93661, 2021. doi: 10.1109/ACCESS.2021.3093421.

[28] A. Al-Shawabka, P. Pietraski, S. B. Pattar, F. Restuccia, and T. Melodia, "DeepLoRa: Fingerprinting LoRa Devices at Scale Through Deep Learning and Data Augmentation," In *Proceedings of the Twenty-second International Symposium on Theory, Algorithmic Foundations, and Protocol Design for Mobile Networks and Mobile Computing*, Jul. 2021, pp. 251–260. doi: 10.1145/3466772.3467054.

[29] S. Cui and I. Joe, "Collision Prediction for a Low Power Wide Area Network Using Deep Learning Methods," *J. Commun. Networks*, vol. 22, no. 3, pp. 205–214, Jun. 2020. doi: 10.1109/JCN.2020.000017.

[30] A. Tsanousa et al., "A Review of Multisensor Data Fusion Solutions in Smart Manufacturing: Systems and Trends," *Sensors*, vol. 22, no. 5, p. 1734, Feb. 2022. doi: 10.3390/s22051734.

[31] L. Gong et al., "An IoT-Based Intelligent Irrigation System with Data Fusion and a Self-Powered Wide-Area Network," *J. Ind. Inf. Integr.*, vol. 29, no. May, p. 100367, Sep. 2022. doi: 10.1016/j.jii.2022.100367.

[32] M. Mahmoud, A. Ashraf Ateya, A. Muthanna, A. Zaghloul, R. Kirichek, and A. Koucheryavy, "Distributed Edge Computing to Assist LPWAN: Fog-MEC Model," In *The 5th International Conference on Future Networks & Distributed Systems*, Dec. 2021, pp. 587–594. doi: 10.1145/3508072.3508192.

[33] S. Wang and L. Yang, "Securing Dynamic Service Function Chain Orchestration in EC-IoT Using Federated Learning," *Sensors*, vol. 22, no. 23, p. 9041, Nov. 2022. doi: 10.3390/s22239041.

[34] J. Gillespie et al., "Real-Time Anomaly Detection in Cold Chain Transportation Using IoT Technology," *Sustainability*, vol. 15, no. 3, p. 2255, Jan. 2023. doi: 10.3390/su15032255.

[35] S. Bhanji, H. Shotz, S. Tadanki, Y. Miloudi, and P. Warren, "Advanced Enterprise Asset Management Systems: Improve Predictive Maintenance and Asset Performance by Leveraging Industry 4.0 and the Internet of Things (IoT)," In *2021 Joint Rail Conference,* Apr. 2021, pp. 1–10. doi: 10.1115/JRC2021-58346.

[36] M. Borova, M. Prauzek, J. Konecny, and K. Gaiova, "A Performance Analysis of Edge Computing Compression Methods for Environmental Monitoring Nodes with LoRaWAN Communications," *IFAC-PapersOnLine*, vol. 55, no. 4, pp. 387–392, 2022. doi: 10.1016/j.ifacol.2022.06.064.

[37] A. Bernard, A. Dridi, M. Marot, H. Afifi, and S. Balakrichenan, "Embedding ML Algorithms onto LPWAN Sensors for Compressed Communications," In *2021 IEEE 32nd Annual International Symposium on Personal, Indoor and Mobile Radio Communications (PIMRC)*, Sep. 2021, vol. 2021-Sept, pp. 1539–1545. doi: 10.1109/PIMRC50174.2021.9569714.

[38] F. Sanchez-Sutil and A. Cano-Ortega, "Smart Regulation and Efficiency Energy System for Street Lighting with LoRa LPWAN," *Sustain. Cities Soc.*, vol. 70, p. 102912, Jul. 2021. doi: 10.1016/J.SCS.2021.102912.

[39] M. Nowak, P. Derbis, K. Kurowski, R. Różycki, and G. Waligóra, "LPWAN Networks for Energy Meters Reading and Monitoring Power Supply Network in Intelligent Buildings," *Energies*, vol. 14, no. 23, p. 7924, Nov. 2021. doi: 10.3390/en14237924.

[40] M. S. Danladi and M. Baykara, "Design and Implementation of Temperature and Humidity Monitoring System Using LPWAN Technology," *Ingénierie des systèmes d Inf.*, vol. 27, no. 4, pp. 521–529, Aug. 2022. doi: 10.18280/isi.270401.

[41] C. Maraveas, D. Piromalis, K. G. Arvanitis, T. Bartzanas, and D. Loukatos, "Applications of IoT for Optimized Greenhouse Environment and Resources Management," *Comput. Electron. Agric.*, vol. 198, no. April, p. 106993, Jul. 2022. doi: 10.1016/j.compag.2022.106993.

[42] E. M. Olalla, H. D. Cadena-Lema, H. M. Domínguez Limaico, J. C. Nogales-Romero, M. Zambrano, and C. Vásquez Ayala, "Irrigation Control System Using Machine Learning Techniques Applied to Precision Agriculture (Internet of Farm Things IoFT)," In *I+D for Smart Cities and Industry*, 2023, vol. 512, pp. 329–342. doi: 10.1007/978-3-031-11295-9_24.

Chapter 9

SDN-based cloud computing secure communication in IoT systems

Muhammad Farhan Siddiqui
Karachi Institute of Economics and Technology (KIET)

Waqas Ahmed Siddique
Millennium Institute of Technology & Entrepreneurship (MiTE)

Awais Khan Jumani
School of Electronic and Information Engineering,
South China University of Technology
ILMA University

Muhammad Arabi Tayyab
Karachi Institute of Economics and Technology (KIET)

9.1 INTRODUCTION

The Internet of Things (IoT) is an emerging innovation that has the ability to completely transform a wide range of sectors. IoT endpoints are prey for hackers because of their growing quantity, though. Current security practices may be insufficient to protect IoT devices from attack. This is because IoT devices are often poorly secured, and they often use outdated or insecure protocols. In addition, the centralized network architecture of SDN makes it difficult to implement complex network-based attack detection solutions [1]. To address these challenges, new security solutions are needed for SDN-based cloud computing architecture for IoT devices. These solutions should be expandable and scalable, so that they can be used to protect a wide variety of IoT devices and applications [2]. Some of the key challenges in implementing network-based attack detection solutions for IoT devices in an SDN environment include:

- **The heterogeneity of IoT devices:** IoT devices come in a wide variety of shapes, sizes, and capabilities. This makes it difficult to develop a single security solution that can protect all IoT devices.

DOI: 10.1201/9781003426974-9

- **The limited resources of IoT devices:** Many IoT devices have finite resources, like memory and computational power. This makes it difficult to implement complicated security solutions on these devices.
- **The complexity of SDN networks:** SDN networks are complex and dynamic. This makes it difficult to design and implement security solutions that can keep up with the changes in the network [3].

Despite these challenges, there are a number of promising approaches to securing IoT devices in an SDN environment [4]. These approaches include:

- **Using machine learning to detect anomalous traffic patterns:** Machine learning can be used to identify patterns in network traffic that are indicative of an attack. This can be used to detect attacks that are not known to the security solution.
- **Using SDN to isolate IoT devices from each other:** SDN can be used to isolate IoT devices from each other. This can help prevent attacks from spreading from one device to another.
- **Using SDN to enforce security policies:** SDN can be used to enforce security policies on IoT devices. This can help to reduce illegitimate access to devices and data.

The security of IoT devices is a key factor that needs to be overlooked. By developing new security solutions for SDN-based cloud computing architecture for IoT devices, we can help to protect these devices from attack and keep our data safe [5].

Security in IoT relates to the security approaches adopted in secure Internet-connected or network-based devices. The term IoT is fairly broad, and as technology advances, the term has only become wider. Almost every electronic gadget can communicate in some form with the Internet devices. IoT security refers to the methods and technology used to prevent these devices from being hacked. Ironically, the fundamental adoption of IoT devices exposes them to hackers. IoT security is significantly more thorough due to the broad reach of IoT. As a result, IoT security currently encompasses a wide range of approaches. IoT security now encompasses everything. These are just a handful of the techniques that IT administrators may use to encounter the increasing number of threats of cybercrime and cyberterrorism stemming from unsecured IoT devices. However, concerns about the protection and safety of the IoT still exist since this equipment is more susceptible to cybersecurity hazards and dangers, and adversaries are always looking for novel techniques to take advantage of the unprotected and unencrypted gadgets that are made accessible to the general audience [6]. Furthermore, IoT gadgets are connected to cloud resources over the world wide web to provide notifications and obtain suggestions for wise activities that IoT unit owners ought to perform. As a result, the existence of IoT peripherals complicates network management and maintenance.

The IoT attack vector has the potential to result in successful security violations. For the protection of essential possessions and facts, effective network safety defenses are needed.

9.2 ROLE OF SDN IN CLOUD COMPUTING

SDN has grown into a crucial domain for network architecture and structure to overcome some issues related to traditional networks. SDN is changing the environment of present-day networks by providing a more flexible and programmable architecture. SDN has a centralized architecture with two different planes, i.e., control plane and data plane. SDN has expanded new research domains for and appears to be a blessing; however, it has some critical issues and challenges including bandwidth utilization, Quality of Service (QoS), and security [7]. SDN has the advantage of centralized network resource management. The SDN controller may be used to manage network resources and monitor network traffic, both of which can improve network security. OpenFlow protocol and custom SDN controller applications have allowed SDN to manage not just standard protocol but also network traffic produced by IoT devices. While SDN may be useful in an IoT environment for tasks like routing management, resource management risk identification and reduction, transportation analysis, and supervisors, it also faces additional risks due to the proliferation of connected devices.

9.3 CLOUD COMPUTING AND IoT SYSTEMS

Through the use of web-based products and services, cloud computing enables individuals to do computer-related tasks. The IoT and cloud computing are now intimately connected caused by the widespread use of IoT combined with cloud innovations. Certain gadgets already exist in the near future and offer a number of benefits [8,9]. Enormous amounts of information have become challenging to store, manipulate, and make available as a consequence of the rapid evolution of innovation. IoT and cloud technology collaboration has produced a tremendous amount of development. It will be possible to combine new tracking solutions with enhanced handling of tactile information channels. Information from sensors, for instance, may be sent to the cloud and stored there to be used for smart analysis and engagement by different gadgets in the future. The goal is to transform information into discoveries that will lead to proceeding that is both efficient and fruitful. The IoT and cloud computing work together to improve everyday tasks' productivity.

Whereas the IoT produces a lot of statistics, cloud computing focuses on creating a pathway for that knowledge to get where it needs to go. Four advantages of cloud computing include:

- Maintains funds since you just invest in the assets you employ; the wider the range, the higher the discounts; there is no obligation to make educated guesses regarding the physical ability needs. Platforms can be implemented globally in a matter of minutes.
- Flexibility and speed in offering resources to developers

As a result, IoT cloud computing works together to preserve IoT information and offer convenient availability when requested. It is crucial to emphasize that cloud computing provides a straightforward method for sending the enormous amount of information produced by IoT across the Internet.

9.3.1 Cloud computing architecture for IoT systems

The architecture of cloud computing combines service orientation with event-driven programming. Cloud computing architecture refers to the system of interconnected parts that make up the cloud. Common components include a front-end platform (desktop, mobile, or thin client), a back-end platform (servers or storage), cloud-based delivery, and a network (Internet, intranet, or intercloud). Together, they comprise the building blocks of cloud computing. Figure 9.1 describes the basic components of cloud computing architecture.

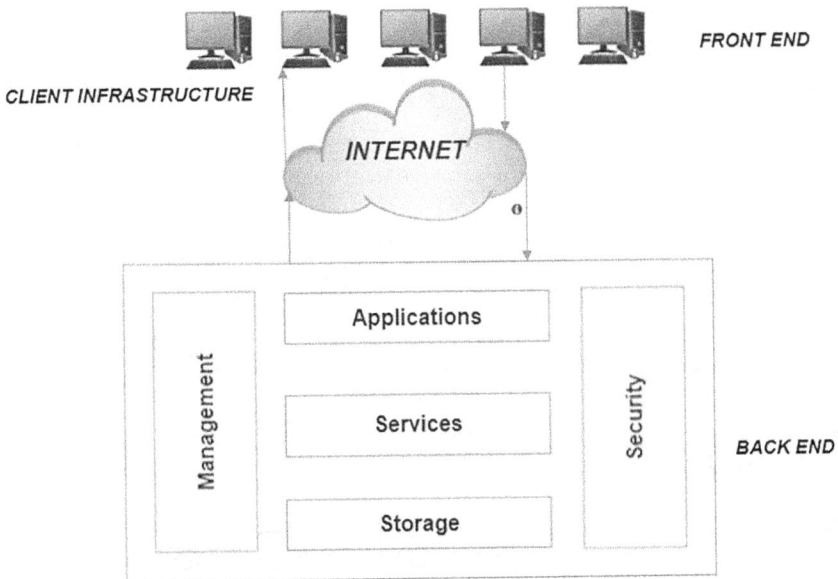

Figure 9.1 Cloud computing architecture for IoT systems.

- **Front end:** The client combines with the front end of the system. It consists of the interfaces of the clients and applications that are necessary to take advantage of cloud computing services. Other front-end components include web servers (such as Chrome, Firefox, and Internet Explorer).
- **Back end:** The service provider utilizes the back end of the system. It is responsible for managing all of the resources that are necessary to provide cloud computing services. It includes a substantial amount of data for storage, various security methods, virtual machines, implementation models, server software, traffic management systems, and other components.

9.4 CLOUD SERVICE MODELS FOR IoT SYSTEMS

9.4.1 Infrastructure as a Service (IaaS)

Hardware as a Service (HaaS) IaaS is also known as (HaaS). It is a computer infrastructure that is connected to the Internet. The essential benefit of using IaaS is that it relieves customers of the cost and inconvenience of maintaining traditional servers. The following traits apply to IaaS:

- Assets are accessible as a service,
- Offerings are extremely scalable,
- Evolving, adaptable, and administration activities are automated as well.

9.4.2 Platform as a Service (PaaS)

Developers may create, assess, run, and handle systems thanks to the PaaS cloud computing framework. The aforementioned traits apply to PaaS:

- More than one person can utilize identical development software.
- Links to archives and Internet resources.
- Assets may be easily adjusted up or down according to virtualization technologies to match the demands of the organization.
- Compatibility with many languages and frameworks.
- Allows you to "Auto-scale" the display.

9.4.3 Software as a Service (SaaS)

SaaS is often referred to as an "on-demand application". The applications included in this program are hosted by a cloud infrastructure vendor. By connecting to the Internet and utilizing a web browser, customers may retrieve these programs. The following traits apply to SaaS:

- Controlled from a single place
- Located on a remote server.
- Internet-accessible
- Both software and hardware modifications are not the responsibility of customers. Automated modifications are performed regularly.
- Each usage of the functions incurs a fee.

9.5 CLOUD IMPLEMENTATION MODELS

9.5.1 Public cloud

This form of cloud is owned and operated by many organizations. These assets, networks, and users number hundreds of thousands of individuals and organizations. Some established public cloud providers are Google, Amazon, and Microsoft. Service distribution, copyright identification, collaborative access administration, and cloud data security from hackers are the main issues with this type of cloud [9].

9.5.2 Personal cloud

This type of cloud is frequently controlled by a particular company and was created specially to satisfy the requirements of a particular enterprise. Using exclusive cloud archiving, businesses may have greater authority over what is stored (which may be subject to regulatory compliance obligations) [9,10]. It may be maintained and hosted either internally or outside. This information might also include medical records, corporate secrets, or other sensitive data. The infrastructure is owned and operated by the same entity.

9.5.3 Hybrid cloud

A handful of global clouds are connected to a personal cloud using the so-called hybrid cloud strategy. Numerous cloud infrastructures with flexible task management are thereby connected and systematically managed. For instance, a company can manage privacy across the two clouds while maintaining valuable information in the internal cloud and general knowledge in the open cloud. Public cloud protection is viewed as less reliable than mixed cloud protection [11,12].

9.6 CHALLENGES IN CLOUD COMPUTING FOR IoT SYSTEMS

9.6.1 Less encryption

Encryption, although being an effective method for preventing hackers from accessing data, is also one of the most significant obstacles to overcome in terms of IoT security. The storage capacity and processing capability of these drives are comparable to that of a conventional computer. As a consequence of this, there has been a rise in the number of assaults in which cybercriminals simply manipulate the algorithms that are designed to provide protection.

9.6.2 Insufficient testing and updating

With the growing number of IoT devices, IoT producers are more eager to create and launch their product as soon as feasible while maintaining security. Most of these gadgets and IoT products are not adequately tested and upgraded, leaving them exposed to hackers and other security threats.

9.6.3 The dangers of predefined credentials and brute force attacks

By virtue of insecure configurations and authentication information, practically all IoT equipment is susceptible to password breaking and brute force assaults. Any company that employs manufacturer-preset passwords on its gadgets exposes its operations, inventory, clients, and sensitive data to a brute force assault.

9.6.4 Ransomware and IoT spyware

The cost rises as more gadgets are produced. While retaining control over a consumer's crucial documents and facts, ransomware is able to effectively block individuals out of a range of gadgets and networks.

9.6.5 Insufficient equipment protection

A lack of suitable safeguards from online assaults, security breaches, theft of information, and unauthorized possession of technological equipment including computers, smartphones, and IoT products is described as poor gadget protection. This might occur as a result of out-of-date applications, passwords that are insecure, unfixed flaws, an absence of data encryption, or additional safety problems. Firmware must be regularly upgraded, and robust safety protocols must be put in place to guarantee the protection and confidentiality of private information held on these gadgets. Numerous IoT gadgets lack security and are susceptible to hacking.

9.6.6 Vulnerability to network attacks

The susceptibility of a network, system, or device to being compromised or exploited by cybercriminals is what is meant when we talk about its vulnerability to network assaults. This may occur as a result of flaws in the design of the network, software that has not been patched, ineffective password management, or a lack of adequate security procedures. Attacks on networks may lead to the theft of data, the disclosure of private information, the suspension of services, and financial loss. It is vital to educate individuals on safe online behaviors and to establish effective security measures such as firewalls, encryption, and regular software updates in order to reduce the likelihood of being vulnerable to assaults on a network. As a result of their dependence on networks, IoT devices are susceptible to assaults known as denial of service (DoS).

9.6.7 Firmware weaknesses

Software bugs are holes or vulnerable spots in the coding of a program that a hacker might use to obtain access without authorization, acquire confidential data, or carry out undesirable operations. The usage of outdated or incompatible programs as well as mistakes or flaws established throughout the building phase can result in program risks. Intruders can take advantage of these flaws to take over a machine, put spyware on it, or grab confidential information. Program authors must utilize appropriate programming techniques, and consumers must maintain their program updated and set up correctly to limit the possibility of bugs. Additionally, organizations and people should utilize strong defensive procedures like firewalls, antivirus software, and intrusion detection systems to safeguard themselves against possible attacks. IoT devices commonly include software weaknesses that attackers might exploit to gain access to devices and networks. To solve these problems, it is critical to adopt security mechanisms like encryption, secure authentication, and software upgrades to guarantee that IoT devices and systems operate safely and securely.

9.6.8 Software-Defined Networking (SDN)

SDN has evolved into an important topic for network architecture and structure in order to solve various issues associated with conventional networks. SDN is transforming the network environment by enabling more flexible and programmable architecture. SDN features a centralized design with two distinct planes, the control plane and the data plane. SDN has opened up new research topics and looks to be a boon; however, it does have certain key concerns and challenges, such as bandwidth usage, QoS, and security.

9.6.9 SDN architecture

In SDN architecture, the controller is responsible for all the responsibilities including all the communication between the northbound and

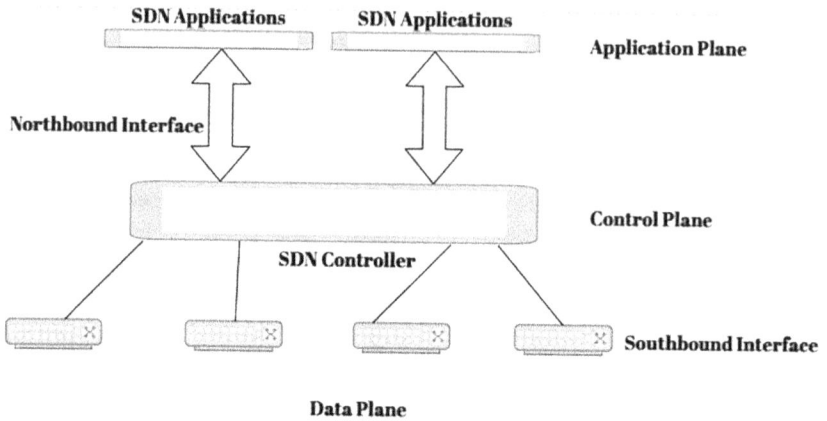

Figure 9.2 SDN architecture.

southbound APIs. The control plane, data plane, and application plane are the three main architectural planes that make up the SDN. The control plane is in charge of network administration, traffic management, and traffic engineering. It is the SDN's most crucial plane. The data plane is made up of components that make up the underlying network used to transmit network traffic. Northbound APIs are used to connect the control and application planes, while the command and information layers communicate through southbound connections. The controller, which interlinks the data plane through the southbound APIs and with applications through the northbound APIs, is the most delicate component of the entire architecture. Additionally, connecting distributed controllers is accomplished by the east-/westbound interface. The network may be dynamically programmed by the SDN controller. The centralized controller also monitors and gathers existing environment condition and data configuration, as well as packet and flow-granularity data, giving it a comprehensive perspective of the network. Figure 9.2 elaborates the basic SDN architecture.

9.7 ADVANTAGES OF SDN FOR IoT SYSTEMS

- Centralized network provisioning: By giving users an integrated picture of the entire system, SDN supports the unification of network administration. By installing both digital and traditional network components from a single location, SDN may also increase resource provision and flexibility.
- Lower operating costs: SDN can assist in reducing operating expenses by providing efficient administration, increased server usage, and enhanced virtualization management. Many routine

network management chores and difficulties may be automated and consolidated.

- **Holistic enterprise management:** SDN allows you to change the configuration of your network devices while maintaining the integrity of your present network. Furthermore, unlike the Simple Network Administration Protocol (SNMP), the administration of physical and virtual switches and network devices is handled by a central controller.
- **More granular security:** By providing a centralized command unit to administer safety and regulatory details for your corporate network, the SDN controller quickly proves to be beneficial for the network itself.
- **Hardware savings and reduced capital expenditures:** Following the SDN controller's instructions, older gear may be reused, while less expensive hardware can be put to best benefit. This procedure converts a new gadget into a white box.
- **Consistent and timely content delivery:** One of the most significant advantages of SDN is the ability to alter data traffic. If you can direct and automate data traffic, it is critical to have high-quality service for Voice over Internet Protocol (VoIP) and multimedia communications. SDN can also assist you in streaming higher-quality videos.

9.8 SDN APPLICATIONS IN IoT SYSTEMS

Software-defined networking (SDN) makes networks more responsive and agile. IoT networking is an architecture that readily abstracts several distinct network layers. SDN attempts to enhance network control by allowing businesses and service providers to adapt swiftly to changing business needs.

9.8.1 Home networking

SDN provide OpenFlow-based system for monitoring and managing users, as well as controlling using "usage caps" or a limited amount of knowledge for every individual or equipment to use the web. The system provides network resource visibility and access management depending on person, enabling consumers to switch broadband bandwidth depending on the person, category, gadgets, implementation, or point in the day.

9.8.2 Security

The controller may analyze and approve connections and the movement of the traffic in the system using OpenFlow-based technology. The host's kernel has a protection component (tagger) that the consumer cannot influence.

This module marks network connections that wish to deliver data (processes, files, etc.). At the commencement of the transmission, these tags are transmitted to the controller (arbiter). The controller examines the tagger and decides whether to approve or refuse the connection based on its regulations. The necessary flow tables are put in the switch after the connection is allowed [8].

9.8.3 Network virtualization

Network virtualization is analogous to OS virtualization in that several data operating and processing systems may share hardware gadgets. That is, the goal of network virtualization is for numerous virtual networks to share identical facilities, but every single has a different topology and navigation algorithm.

9.8.4 Multimedia

SDN enables multimedia management duties to be optimized. This design consists of a pair of components: the Path Assignment Function (PAF), which periodically updates the overall network layout, and the QoS Balancing and Optimization Function (QMOF), which analyses the different interactive media properties to identify the optimum setup for the given path. In the event that the quality of the links deteriorates, the system automatically updates the route parameters while keeping the users' preferences in mind.

9.8.5 Reliability and recovery

One of the most typical issues in conventional networks is the difficulty in recovering from a connection breakdown. The convergence time is influenced by the node's restricted ability to recalculate the path. SDN's global view allows for the customization of recovery methods.

9.9 THREATS AND CHALLENGES TO SECURE COMMUNICATION IN IoT SYSTEMS

9.9.1 Software and firmware vulnerabilities

Because many smart devices are resource-constrained and have limited processing capabilities, it is difficult to guarantee the security of IoT systems. This is one of the reasons why ensuring the security of IoT systems is so challenging. As a consequence of this, they are unable to carry out comprehensive, resource-intensive security activities, making them more susceptible than devices that are not connected to the IoT. The following are some of the reasons why many IoT systems are susceptible to security attacks:

- Insufficient computing power for effective built-in encryption.
- Access management issues with IoT platforms.
- Inadequate funding for thoroughly evaluating and enhancing software integrity.
- The absence of frequent repairs and upgrades as a result of IoT equipment technological constraints and budgetary constraints.
- End users might not upgrade their gadgets, which would prevent vulnerabilities from being fixed.
- Over time, devices that date back may cease to be capable of receiving system upgrades.
- Inadequate defense against physical assaults: a perpetrator only has to get sufficiently close to insert a chip or use radio frequencies to compromise the equipment.

9.9.2 Insecure communications

Unsecured lines of communication expose users to several dangers, one of the most serious of which is the possibility of a man-in-the-middle attack (MitM). Hackers may easily gain control of your device by using MitM attacks if the device does not utilize strong encryption and authentication protocols. These attacks can be used to compromise an update operation and take control of your device. Attackers are even able to install malware or interfere with the functioning of a device. Even if your device has not been hacked by a MitM attack, the data that it exchanges with other devices and systems and sends in plain text conversations might still be collected by hackers.

9.9.3 Data leaks from IoT systems

We have shown in the past that hackers are able to acquire the information that is being processed by your IoT device by simply intercepting messages that are not encrypted. There is a possibility that this contains private information about you, such as your location, bank account details, or medical history. However, exploiting insecure connections isn't the only way that attackers might get their hands on vital information; they also have other options. Because of this, cloud-hosted services are susceptible to attacks from the outside world. All of the data is sent and kept on the cloud. As a consequence of this, it is possible for data to be stolen from devices as well as the cloud environments to which the devices are tied. It's possible that data might be leaked from your IoT devices due to third-party services.

9.9.4 Malware risks

Intruders may change an IoT mechanism's functionality, acquire user facts, and launch other attacks if their efforts are fortunate in inserting viruses into it. In addition, specific gadgets could be pre-infected with malware if vendors do not supply enough application security. Some firms have already discovered strategies to combat the most well-known IoT-targeted malware.

9.10 SECURE COMMUNICATION PROTOCOLS FOR IoT SYSTEMS

9.10.1 MQTT (Message Queuing Telemetry Transport)

Message Queuing Telemetry Transport, more often known as MQTT, is a messaging transport protocol that facilitates communication between clients and servers. TCP/IP or alternative protocols may be used for MQTT in as much as they provide the unaltered, necessary two-way interactions [13].

- The protocol is uncomplicated and very lightweight, and it allows for the simple and quick transfer of data.
- MQTT was developed with restricted devices in mind, as well as networks that have poor bandwidth, are unreliable, or have excessive latency.
- Using fewer data packets assures that there will be less demand placed on the network.
- Because flawless power management defends the less usage of the battery of the linked equipment, this technology is ideal for mobile phones and wearables, which have a need for the usage of the battery to be kept to a minimum.

MQTT security is separated into three layers: the first one is the layer of the network, the second one is the layer named transport, and the third layer is application. Each layer protects against a certain form of assault. MQTT is a low-level rule and regulation that only explains the working on a few security measures [13]. Other security standards typically used in MQTT implementations include SSL/TLS for transport encryption, VPN at the network level for a physically secure network, and the usage of username/password. Data packets contain a client identification that is used to authenticate devices at the application level.

9.10.2 CoAP (Constraint Application Protocol)

A web exchange framework called CoAP (Constraint Application Protocol) was created for restricted equipment (such as microcontrollers) and low-powered or lossy systems. It is also a popular protocol for securing IoT applications [13,14].

- CoAP follows the REST paradigm in much the same way as HTTP does. Clients make use of protocols like GET, PUT, POST, and DELETE to make the availability of the entities that can be easily checked through several URLs.
- Since CoAP was developed to run on microcontrollers, it is an ideal candidate for the IoT, which calls for millions of low-cost nodes to be connected.
- CoAP makes very little use of the resources available on the device as well as those available on the network. It does not need a complicated transport stack since it relies on UDP over IP.
- Due to the fact that its factory setting of Datagram Transport Layer Security (DTLS) rules and regulations is comparable to the Key of the RSA of 3072-bit, CoAP is considered to be one of the most secure protocols.

CoAP transports information via UDP (User Datagram Model) and so it has faith in the UDP surveillance characteristics to safeguard the required data. CoAP used the Datagram TLS over UDP.

9.10.3 Datagram Transport Layer Security

The DTLS protocol is a security protocol for the IoT that is meant to safeguard data transfer between data-gram-based applications. It uses the TLS (transport layer security) protocol and offers the same degree of security.

- To address the problem of packet loss, DTLS implements a retransmission timeout in its protocol. In the event that the timer runs out before the client even receives the acceptance notification from the server side, the user will send the information again.
- The problem of asking again and again may be circumvented by assigning a unique data number also known as the sequence number to every message. This assists in assessing whether or not the subsequent message that is received is in sequence. If it is not in the correct order, it will be placed in a queue and attended to when the appropriate number in the sequence is reached.
- The DTLS protocol is unstable and does not ensure that data will be sent, not even for payload information.

9.10.4 ZigBee

ZigBee is thought to be a cutting-edge technology for providing security for IoT devices and applications. The integrated machinery having less power, such as radio systems, helps to establish effective end-to-end, or, in other words, machine-to-machine transmission from 10 to 100 m away. It is a low-cost, open-source wireless technology [15]. There are two models of the security which are appropriate with ZigBee:

9.10.5 The networked centrally managed privacy

This not only makes the system more secure but also makes it more complex since in it a third equipment service is taken. This third-party system is known as the Trust Centers. Trust Centers are operational software that runs on devices that are trusted by other devices in the ZigBee network. This makes the system more secure, but it also makes it more complex. A centralized network is created by the Trust Centre, which also sets up and authorizes every device that wishes to join it by allocating it a special TCLK (TC Link Key) [15].

9.10.6 Platform for distributed protection

DSN is easier but less trustworthy than CSN since it lacks a Core Node or Trust Centre. It is possible for each gateway to create a distributed networking on its own accord. The network key is the only thing a node obtains when it enters the network [15,16].

9.11 SECURITY MECHANISMS FOR SECURE COMMUNICATION IN IoT SYSTEMS

An essential privacy technique for integrated microcontrollers in the IoT products ensures safe interaction. Encoding information that is in circulation during encrypted interaction prevents hackers from accessing or changing it. The knowledge delivered in and out of the cloud, information exchanged back and forth among gadgets, may be protected via secure communication. Secure protocols, such as TLS and DTLS, are commonly used to implement secure communication. These protocols offer another degree of protection by authenticating devices and encrypting data. This guarantees that only authorized devices have access to the data and that it remains secure while in transit. It is also critical to have secure communication routes in addition to secure protocols. VPNs and encrypted Wi-Fi networks, for example, give an extra degree of protection by encrypting data in transit. This guarantees that the data is safe and that bad actors cannot access or modify it.

9.11.1 Secure storage

Encrypted memory is another important safety element for integrated microcontrollers in IoT products. Protected preservation is a method of keeping content in a safe area of an instrument, such as a private component or a Verified Platform Module (TPM). This shields the information from malicious parties and stops unauthorized editing or viewing. Credentials and other delicate data can be kept in secure cabinets. To guarantee that only those authorized changes get installed, it is additionally possible to save bios and system versions.

9.11.2 Secure updates

Reliable patches are yet another critical safety precaution for implanted microcontrollers in IoT systems. Confidential repairs refer to the method of securely upgrading an item's OS and architecture. This ensures that the gadget is capable of being updated with authorized versions and prevents hackers from altering the operating system to get into the machine [17,18].

9.12 SDN-BASED SECURE COMMUNICATION IN IoT SYSTEMS

The IoT landscape is divided into three layers: The IoT device layer, the IoT network platform, and the IoT application. This overview covers everything from equipment to pertinent IoT applications [19].

- **IoT device layer:** This layer contains all devices that may interact with physical surroundings through the use of identification and sensing capabilities. This layer is the primary source of data collecting for analysis inside the network platform. RFID and wireless sensor networks (WSN) are two technologies that are commonly used at this layer [19].
- **IoT network layer:** This layer, known as the landscape's heart, provides a variety of services like network administration, communications, and security. Because of the rising interconnectedness of smart items, cloud technologies have been at the forefront of this layer. Low latency is a critical requirement for this layer [19].
- **IoT application layer:** This layer contains all of the application modules necessary to provide the end user with the desired service. IoT systems vary widely in order to meet various business logic needs [19].

9.13 SDN-BASED SECURE COMMUNICATION ARCHITECTURE FOR IoT SYSTEMS

The current network paradigm is proving inadequate for safeguarding IoT systems, leading to heterogeneity and restricted scalability. SDN has been examined by researchers as a means of enhancing IoT bandwidth and flexibility. The primary benefit of SDN is the separation of the control plane and the data plane. The SDN controller makes decisions, whereas the switches are only forwarding components. This aids network administration and scalability. Another advantage of utilizing SDN is the ability to control flows dynamically. If there is ambiguity about future forwarding choices, switches can forward the packet to the controller. SDN applications can communicate their network requirements to the SDN controller in this manner, allowing for the adjustment or inclusion of new flow rules. SDN makes network services more agile and versatile since they can be installed and configured automatically. With these added benefits, SDN has seen widespread adoption in IoT systems [20,21].

9.14 SDN-BASED SECURITY MECHANISMS FOR IoT SYSTEMS

- **Traffic isolation:** SDN may be used to allow various network traffic to be routed over the same physical infrastructure while maintaining the needed isolation. This capability can significantly decrease the spread and impact of security assaults across several network domains. This is a critical component in the IoT system because sensitive processes may rely on data supplied by other items. On top of the SDN controller, we can also create an SDN-based IoT access control application. As a result, all incoming and outgoing connections to the IoT system may be validated in accordance with the security regulations. Due to SDN's dynamic programmability, hostile traffic may be dynamically prevented, resulting in a quicker response security mechanism [19].
- **Dynamic flow control:** Because SDN controllers may dynamically install and update forwarding rules, network applications can incorporate appropriate security methods. This is feasible because the data plane and control plane have been decoupled. When a switch lacks a flow rule to determine how to process an incoming packet, a request can be routed to the controller, who responds with an appropriate flow rule. This feature can offer a dynamic access control function, which aids in network administration and network privacy.
- **Network programmability:** The network application software can regulate data forwarding in the SDN network. This increases SDN's flexibility in enabling additional security functionalities. To manage offloading the greater SDN programmability, an Enterprise Centric Offloading System has been suggested. SDN programmability enables remote administration and easy application updates as new dangers surface.

- **Centralized network monitoring:** The SDN controller has insight into the data planes under its control. It can gather network status information via the control plane by sending statistics inquiry messages to the switches. The SDN controller can offer the applications with information such as status reports and flow request messages. This method aids in the design and implementation of methods for implementing anomaly network analysis and detection in national networks.

9.15 CLOUD COMPUTING SECURITY FOR IoT SYSTEMS

9.15.1 Insecure interfaces and APIs

The program Programming Interface (API) allows communication between the program and the server. API security is critical to cloud security. If there is a shaky set of APIs, security concerns are more likely to occur [6,22].

9.15.2 Misuse of cloud services

Abuse of cloud services happens when customers of cloud platforms violate the terms of their agreements with such platforms. Attacks such as brute force, Trojans, SQL injection, botnets, phishing, and DoS attacks are almost guaranteed to be carried out by malicious attackers. As a result of their inability to launch or thwart attacks, service providers are unable to determine whether or not attacks were launched on their networks [6].

9.15.3 Malicious intruders

These individuals operate on the cloud platform, have access to the data and resources of users, and are involved in data manipulation. Because cloud computing is extensively used and welcomed across the world, IoT relies on it for data and resource storage. The two computer technologies are vulnerable to security risks, which is a serious concern. It is critical to mitigate security concerns in order to preserve consumer confidence and integrity. Because IoT and cloud computing technologies are continually evolving, genuine security is essential [13–18,23–28].

9.15.4 Denial of Service

The purpose of an attack known as a DoS is to stop cloud users from accessing the data or applications they pay for. When this occurs, the service consumes a substantial portion of the available resources on the system, such as memory, disk space, or bandwidth on the network. If the attackers or hackers are successful in achieving this goal, it will almost always cause a delay in the system, and consumers of cloud services will be unable to use the service as intended after the DoS attack. It is possible for several attackers or attack sources to execute simultaneous DoS attacks, which is referred to as distributed denial of service (DDoS) [6,29,30].

9.15.5 Forwarding device attack

This attack can be carried out with the help of access points and switches using DoS assaults, and it can bring down or temporarily disturb the network system.

9.16 CLOUD COMPUTING SECURITY SOLUTIONS FOR IoT SYSTEMS

9.16.1 Encrypt your data

Encryption serves to safeguard the information by encoding it whenever the data is both at rest and in motion. It is an essential part of the IoT cloud security. Without the appropriate decryption key, the encrypted material is very difficult to decode. The use of such a key, in turn, is restricted to a certain number of individuals. On the other hand, it is not viable to develop a single solution for the encryption of data for the IoT that is applicable to both smart kettles and massive construction equipment. Therefore, IoT cloud security professionals modify encryption techniques in accordance with the kind of environment (hub, edge, storage device, etc.) that they are encrypting [24]. When it comes to the security of IoT and the cloud, safe protocols for the IoT are of the highest significance. Protocols such as MQTT, CoAP, and XMPP, among others, will ensure the safety of your IoT solutions [31].

9.16.2 Leave sensitive data on-premises

Despite the fact that cloud service providers place a high priority on data security, it is nevertheless fraught with danger to keep sensitive information on a public cloud. There are three key clusters of sensitive data that have to be stored locally. The terms "personally identifiable information" (PII), "personal healthcare information" (PHI), and "financial data" are used to refer to these types of data [26,32].

9.16.3 Use DevSecOps approach

When it comes to the security of the IoT and the cloud, the DevSecOps strategy is the method of choice. Within the framework of this methodology, safety precautions are taken throughout each and every step of the development process. As a result of the fact that each developer is responsible for the security of their own code, evaluating security risks has become an essential component of the process of developing software. In addition to this, including security at an earlier stage of development enables the risk of security breaches to be reduced [27,32].

9.16.4 Use of machine learning techniques

ML has emerged as a key technology in the telecommunications industry. One standout feature of the platform has been the integration of ML with SDN. ML is used to bring out important patterns and models from the system's learning set. Essentially, it has two phases. To build a model from the dataset for identifying network abnormalities, the learning phase use ML techniques to learn the sequences. In the decision-making phase, the model makes predictions and takes action based on the training data it has received. The three types of ML techniques are semi-supervised, unsupervised, and supervised. Using labeled input as training data, supervised learning techniques employ the trained model to predict unknown situations. Unsupervised learning attempts to anticipate the data cluster or structure in the dataset using unlabeled data as learning data. When the trained model receives new data, it places it in one of the clusters [33,34]. In the period of 12–18 months, 61% of firms want to engage in IoT hazard evaluation, as reported by Gartner. Consequently, there is a growing demand for experienced cloud safekeeping personnel. As an outcome, to accomplish IoT and cloud protection, many firms choose to work with knowledgeable IT providers [32].

9.16.5 Physical security of the devices

For the purpose of onsite security of the devices Artificial intelligence techniques can be applied including face recognition and biometric systems for the physical security of the devices and data in order to avoid or reduce unauthorized access in data centers [35,36].

9.17 OPEN RESEARCH ISSUES

To make the SDN more secure, researchers have attempted to incorporate additional hardware or software techniques. Network function virtualization (NFV) appears to be a perfect fit since it allows us to create programmable switches that are objects or instances, making network management easier. ML has infiltrated SDN networks, improving intrusion detection systems. ML is being used to identify malware, which has the potential to gravely disrupt IoT equipment. Frameworks are being created to make

IoT systems more programmable. Because IoT systems are notoriously resource-restricted, these frameworks are often lightweight, and they are built in programmer-friendly languages for simple comprehension and greater collaboration.

9.18 CONCLUSION

The IoT ecosystem is constantly changing, drawing a rising number of fraudsters looking to exploit IoT system vulnerabilities to launch harmful attacks on a potentially global scale. Conventional security methods are ineffective in dealing with security concerns in this setting. We explored numerous dangers, responses, and challenges that may emerge while using SDN in IoT systems in this post. Despite the concerns, the future of SDN may be consolidated into a very safe system when integrated with diverse sectors such as ML and data analysis.

REFERENCES

[1] Chaganti, R.; Suliman, W.; Ravi, V.; & Dua, A. (2023). Deep learning approach for SDN-enabled intrusion detection system in IoT networks. *Information*, 14(1), 41.

[2] Siddique, W.A.; Siddiqui, M.F.; & Khan, A. (2020). Controlling and monitoring of industrial parameters through cloud computing and HMI using OPC Data Hub software. *Indian J. Sci. Technol.*, 13(2), 114–126.

[3] Maddikunta, P.K.R.; Gadekallu, T.R.; Kaluri, R.; Srivastava, G.; Parizi, R.M.; & Khan, M.S. (2020). Green communication in IoT networks using a hybrid optimization algorithm. *Comput. Commun.*, 159, 97–107.

[4] Batool, R.; Siddiqui, E.M.F.; Mushtaq, N.; & Ali, W. (2019). Review, analysis of SDN and difficulties in adoption of SD. *J. Inf. Commun. Technol. (JICT)*, 13(1).

[5] Lee, I.; & Lee, K. (2015). The Internet of Things (IoT): applications, investments, and challenges for enterprises. *Bus. Horizons*, 58, 431–440.

[6] Saini, D.K.; Kumar, K.; & Gupta, P. (2022). Security issues in IoT and cloud computing service models with suggested solutions. *Security Commun. Netw.*, 2022, pp. 169–184.

[7] Farhady, H.; Lee, H.; & Nakao, A. (2015). Software-defined networking: a survey. *Comput. Netw.*, 81, 79–95.

[8] Tawalbeh, L.A.; Muheidat, F.; Tawalbeh, M.; & Quwaider, M. (2020). IoT privacy and security: challenges and solutions. *Appl. Sci.*, 10(12), 4102.

[9] Ahmad, W.; Rasool, A.; Javed, A.R.; Baker, T.; & Jalil, Z. (2022). Cyber security in iot-based cloud computing: a comprehensive survey. *Electronics*, 11(1), 16.

[10] Rani, B.K.; Rani, B.P.; & Babu, A.V. (2015) Cloud computing and inter-clouds-types, topologies and research issues. *Procedia Comput. Sci.*, 50, 24–29.

[11] Diaby, T.; & Rad, B.B. (2017). Cloud computing: a review of the concepts and deployment models. *Int. J. Inf. Technol. Comput. Sci.*, 9, 50–58.

[12] Shaikh, A.H.; & Meshram, B. (2021). Security issues in cloud computing. In *Intelligent Computing and Networking*; Springer: Singapore, pp. 63–77.

[13] Gerodimos, A.; Maglaras, L.; Ferrag, M. A.; Ayres, N.; & Kantzavelou, I. (2023). IOT: Communication protocols and security threats. Internet of Things Cyber-Phys. Syst., 3, 1–13

[14] Radiocrafts. "CoAP Constrained Application Protocol." Retrieved from: https://radiocrafts.com/technologies/coap-constrained-application-protocol/ #:~:text=Constrained%20Application%20Protocol%20(CoAP)%20 is,low%20bandwidth%20and%20low%20availability.

[15] Karmokar, N. "Top 5 Security Protocols for the Internet of Things (IoT)." Dhi Technologies. Retrieved from: https://www.dhitechnologies.org/ top-5-security-protocols-for-the-internet-of-things/

[16] Khan, M.O.; Jumani, A.K.; & Farhan, W.A. (2020). Fast delivery, continuously build, testing and deployment with DevOps pipeline techniques on cloud. *Indian J. Sci. Technol.*, 13(5), 552–575.

[17] Shacklett, M.E. (2021). "Reinforce IoT Cloud Security in 6 Steps." Retrieved from: https://www.techtarget.com/iotagenda/tip/Reinforce-IoT-cloud-security -in-6-steps#:~:text=Secure%20network%20protocols%20%2D%2D%20 including,for%20each%20IoT%20device%20individually.

[18] Alibaba Cloud Bao. (2023). "Security Mechanisms for Embedded Microcontrollers in IoT Devices." Retrieved from: https://www.alibaba-cloud.com/tech-news/tutorial/giqt0fpftt-security-mechanisms-for-embed-ded-microcontrollers-in-iot-devices#:~:text=Secure%20communication%20 is%20typically%20implemented,is%20secure%20while%20in%20transit.

[19] Patrick, G. (2022). "SDN for Securing IOT Systems." Dell Technologies Proven Professional Knowledge Sharing. Retrieved from: https://education. dellemc.com/content/dam/dell-emc/documents/en-us/2022KS_Patrick-SDN_ for_Securing_IoT_Systems.pdf.

[20] Laghari, A.A.; Jumani, A.K.; & Laghari, R.A. (2021). Review and state of art of fog computing. In Eugenio Oñate, Michal Kleiber (eds.), *Archives of Computational Methods in Engineering* (pp. 1–13), Springer, Germany.

[21] Uddin, M.; Khalique, A.; Jumani, A.K.; Ullah, S.S.; & Hussain, S. (2021). Next-generation blockchain-enabled virtualized cloud security solutions: review and open challenges. *Electronics*, 10(20), 2493.

[22] Microsoft. (2021). "What Is PaaS? Platform as a Service." Microsoft Azure. Retrieved from: https://azure.microsoft.com/en-in/overview/what-is-paas/#overview.

[23] Microsoft Azure. "IoT Security: An Overview." Retrieved from: https://azure. microsoft.com/en-in/resources/cloud-computing-dictionary/what-is-iot/ security/

[24] Embitel. (2020). "IoT Security - Part 3 of 3: IoT Cloud Security and IoT Application Security." Retrieved from: https://www.embitel.com/blog/embedded-blog/iot-cloud-and-application-security

[25] Nduati, J. (2020). "IoT and Cloud Computing Security Threats." Retrieved from: https://www.section.io/engineering-education/iot-and-cloud-computing-security-threats/.

[26] Jumani, A.K.; Siddique, W.A.; & Laghari, A.A. (2023). Cloud and machine learning based solutions for healthcare and prevention. In *Image Based Computing for Food and Health Analytics: Requirements, Challenges, Solutions and Practices: IBCFHA* (pp. 163–192). Cham: Springer International Publishing.

[27] Jumani, A.K.; Shi, J.; Laghari, A.A.; Hu, Z.; Nabi, A.U.; & Qian, H. *Fog Computing Security: A Review.* Security Privacy, e313.

[28] Siddique, W.A.; Jumani, A.K.; & Laghari, A.A. (2022). Introduction to Internet of Things with Flavor of Blockchain Technology. In *Blockchain* (pp. 51–72). London: Chapman and Hall/CRC.

[29] Boppana, R.V.; Chaganti, R.; & Vedula, V. (2019). Analyzing the vulnerabilities introduced by ddos mitigation techniques for software define networks. In *National Cyber* Summit; Springer: Berlin/Heidelberg, Germany, pp. 169–184.

[30] Catalin, C. (2018). "AWS said it mitigated a 2.3 Tbps DDoS attack, the largest ever IZDNet".

[31] Khan, A.A.; Laghari, A.A.; Shaikh, A.A.; Shaikh, Z.A.; & Jumani, A.K. (2022). Innovation in Multimedia Using IoT Systems. In Rajeev Tiwari, Neelam Duhan, Mamta Mittal, Abhineet Anand, Muhammad Attique Khan (eds.), *Multimedia Computing Systems and Virtual Reality* (p. 171). CRC Press, Boca Raton.

[32] Zabor, K. (2021). "5 Best Practices to Ensure IoT and Cloud Security." Retrieved from: https://www.n-ix.com/best-practices-ensure-iot-cloud-security/.

[33] Ahmad, A.; Harjula, E.; Ylianttila, M.; & Ahmad, I. (2020). Evaluation of machine learning techniques for security in SDN. In *2020 IEEE Globecom Workshops (GC Wkshps)* (pp. 1–6). IEEE.

[34] Jiménez, M.B.; Fernández, D.; Rivadeneira, J.E.; Bellido, L.; & Cárdenas, A. (2021). A survey of the main security issues and solutions for the SDN architecture. *IEEE Access,* 9, 122016–122038.

[35] Siddiqui, M.F.; Siddique, W.A.; Ahmedh, M.; & Jumani, A.K. (2020). Face detection and recognition system for enhancing security measures using artificial intelligence system. *Indian J. Sci. Technol.,* 13(9), 1057–1064.

[36] Zhuang, L.; Jumani, A.K.; & Sbeih, A. (2021). Internet of things-assisted intelligent monitoring model to analyse the physical health condition. *Technol. Health Care,* 29(6), 1277–1290.

Chapter 10

Big data analytics for 6G-enabled massive internet of things

Waqas Ahmed Siddique
Millennium Institute of Technology & Entrepreneurship (MiTE)

Muhammad Farhan Siddiqui
Karachi Institute of Economics and Technology (KIET)

Awais Khan Jumani
South China University of Technology
ILMA University

Waheeduddin Hyder
Millennium Institute of Technology & Entrepreneurship (MiTE)

Abdul Ahad Abro
Ege University

10.1 INTRODUCTION

Fifth-generation cellular infrastructure is replaced by sixth-generation wireless or 6G. The bandwidth and frequency of 6G connections will be considerably greater than those of 5G connections due to their ability to operate at larger frequencies. One of the objectives of the 6G online world is to provide interaction with a 1-μs speed. It's nearly 1,000 periods quicker instead of 1 ms throughput, or 1/1,000th the latency. The growing demand for 6G computing is predicted to enable significant advancements in picturing, appearance technologies, and sense of location. The 6G computational architecture will be capable of determining the ideal location for determining to happen, including choices for information preservation, handling, and exchange. This capability is going to function in concert with the use of artificial intelligence (AI). It's crucial to remember that 6G has not yet become an operational product. The market's requirements for 6G-capable infrastructure solutions are still years away despite some manufacturers engaging in the future-oriented wireless accepted standards [1].

DOI: 10.1201/9781003426974-10

Massive IoT, also known as Massive Machine Type Communications (mMTC) in 5G nomenclature, is a brand-new name for an established concept: using cellular connections to link an enormous amount of battery-powered, mostly unsupervised gadgets in a simple and affordable manner. Massive IoT modules are ideal for detection devices that can be dispersed across large regions since they have long lifetime lives and slow transmission rates. The IoT on an unparalleled scale is referred to as massive IoT. It is a widespread phenomenon that integrates the capacity of numerous detectors, connections, and data analysis to provide fresh approaches to solving issues. The widespread IoT is transforming the manner in which we practice business, from being able to keep track of the condition of water to identifying when a package is on its journey [2].

There are various aspects in which the massive IoT varies in comparison to the traditional IoT. It is obvious that this new communication method links a lot of machines. The phrase has been in circulation since around 2018, but as LPWANs and 5G acquire popularity, its significance has increased. There's further, though. Massive IoT constitutes an ecosystem that combines big data information analysis, cloud and edge processing, and AI to make enormous amounts of knowledge more useful and approachable to enterprises. IoT devices are expected to produce 73.1 ZB worth of information by 2025 (IDC). That is around four times the 18.3 ZB produced in 2019. Walmart, for illustration, stated in 2021 that its employees supervise over 1.5 billion IoT equipment communications every day. The firm employs IoT, in particular, to assist in keeping food in freezers and freezers and anticipate the necessity for in-store repair. Therefore, it should come as no shock that the expression "massive" generally encompasses the reality that this kind of system processes large amounts of knowledge at rapidly growing levels; in a nutshell, this is generally the range of big data [2,3].

10.2 MASSIVE IoT AND CRITICAL IoT

IoT is divided into two distinct groups: essential IoT and huge IoT. Essential IoT activities must meet strict standards for accessibility, interruption, and dependability. For instance, activities for interconnected autos, self-driving transportation, industrial software, offsite clinical procedures in medical care, and roadway security (V2X, or "vehicle to anything"). Figure 10.1, Massive IoT, on the contrary conjunction, has many relationships, tiny content quantities, occasional communication, inexpensive gadgets, and strict resource usage restrictions. Some instances involve intelligent meters, devices that are worn, automated buildings, fleet administration, industrial tracking, transportation operations, and agribusiness [2].

Figure 10.1 Traffic-safety use cases are often associated with critical IoT [2].

The following are the main distinctions among huge and essential IoT [2]:

- The ability to scale is not as important for huge IoT as it is for essential IoT.
- Protection is necessary for both big data and essential IoT applications.
- Communication latency for vital IoT applications, reduced latency is essential.

The experience of earlier versions reveals that as IoT develops, 5G will eventually attain its limits and be considered inadequate to handle the majority of such sophisticated services (Figure 10.2). To allow widespread IoT, there must be a significant incentive for sixth-generation technologies (6G) to raise the bar on 5G functionality [2].

Big Data is a buzzword used to describe a sizable compilation of organized and unorganized information that is exceedingly challenging to handle using conventional methods. However, it's crucial to analyze company facts in order to gain knowledge that may be used to guide business choices that are strategic. Data processors employ a variety of technologies to create valuable knowledge from disorganized facts. Each moment that we engage with a piece of innovation, whether they're an intelligent gadget, an application for fitness, or an e-commerce software, we create an enormous number of records that businesses may utilize to keep tabs on our preferences and behaviors. This aids businesses in effectively grasping consumer wants and producing/providing access to users' goods and activities. Big

Figure 10.2 Massive IoT use cases include fleet management, transport logistics, and asset tracking [2].

data is improving our existence as well as influencing them. Details collection and storage methodologies vary. IoT equipment is usually among the primary methods of gathering information. These gadgets contain built-in detectors that gather information about the surroundings. Via the global web, the cloud receives the essential information that has been gathered. IoT gadgets are used by businesses to gather information. IoT appliances save their content in an unprocessed format; therefore, big data analyzes this information at the same time and maintains it utilizing a variety of collection platforms. Consequently, there currently exists a pressing demand for enormous amounts of data in IoT [4].

10.3 IoT BIG DATA PROCESSING OCCURS IN FOUR SEQUENTIAL STEPS

- IoT gadgets produce and preserve an assortment of unprocessed statistics in a big data system.
- A massive data platform is a decentralized networked database that stores a significant quantity of knowledge.
- Analytical techniques like Hadoop MapReduce or Spark are used for evaluating any information that has been recorded.
- Next, create summaries based on the results of the analysis.

Big Data and the IoT enable businesses across a variety of industries to formulate quick, educated judgments that improve their offerings and outcomes. IoT in Big Data research aids firms with knowledge extraction to gain more insightful business decisions. Enhanced business perspectives aid in making selections that have an excellent return on investment. Organizations are transitioning to large data cloud warehousing because it decreases the expenses of implementation owing to a rise in desire for data retention [5]. Big data from IoT enables businesses to

- Evaluation of information
- Highlight patterns in the data.
- Discover unknown data correlations.
- Look for buried data correlations.
- Disclose fresh details.

Big data insights are the sometimes challenging method for sifting through enormous amounts of statistical information to find insights that may assist businesses in making wise judgments about their operations. Big data insights, technologies, and applications may help organizations generate evidence-driven information that can enhance the results of their company operations. Rewards might involve enhanced consumer personalization, increased administrative effectiveness, and greater profitable advertising. These positive aspects over competitors may be achieved with a strong tactic [6].

10.4 CHARACTERISTICS OF BIG DATA IN 6G-ENABLED MIOT

The characteristics of big data that are most frequently mentioned are:

- Quantity
- Diversity
- Truthfulness
- Velocity
- Value

Quantity and diversity, the first two criteria, correspond to the equipment and programming necessities for managing enormous diverse collections of information, while diversity and velocity correspond to the capabilities for real-time computing with enough reliability. Figure 10.3, On the contrary conjunction, multidisciplinary collaboration between academics, businesses, and the global sectors is necessary to extract the most valuable information possible from the complicated massive data streams in wireless IoT platforms [7].

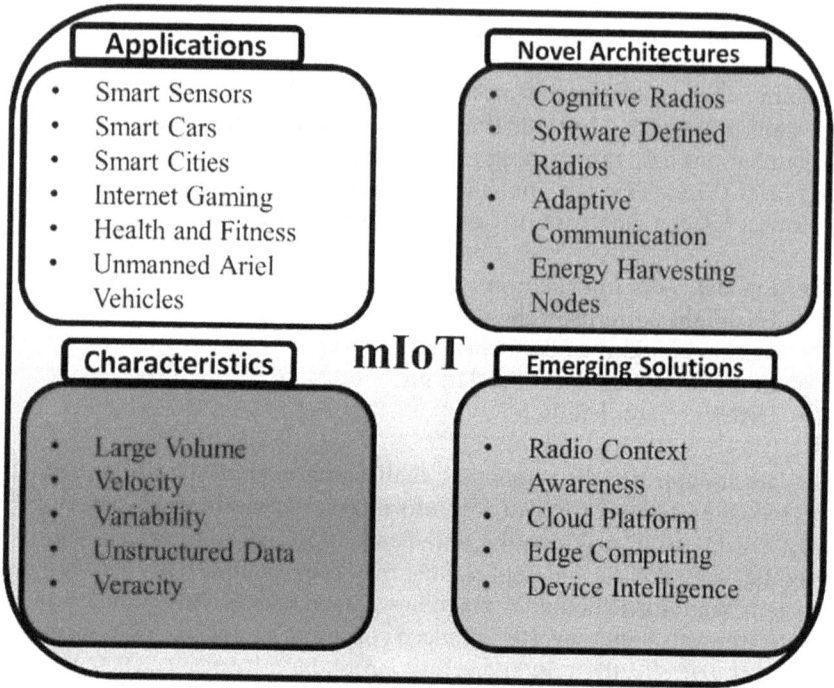

Figure 10.3 Taxonomy of Multimedia-enabled Internet of Things (mIoTs) [8].

10.5 CHALLENGES IN BIG DATA ANALYTICS FOR 6G-ENABLED MIOT

The features of MIoT content are as follows: [9] vast quantity, [10] diverse data kinds, [11] to be created in real-time, and [12] delivering enormous societal and corporate significance. For MIoT, the distinctive qualities provide significant difficulties in big data insights. We list the difficulties under the subsequent headings.

10.5.1 Problems with data collection

The following difficulties arise in gathering information and distribution, which are addressed through data intake.

- Having trouble representing data. MIoT input comes in a variety of formats, heterogeneous frameworks, and measurements. According to Tao et al. (2018), production knowledge may be divided into organized, partially structured, and unorganized categories. One of the main issues in big data processing for MIoT involves how to deal with this organized, partially structured, and unprocessed input.

- Effective data transfer. Because of multiple factors, transferring massive amounts of information to a preservation facility efficiently turns into challenging:
  - High-capacity use because massive data transfer has become a significant constraint for wireless telecommunication platforms.
  - Among the main limitations for numerous wireless industrial infrastructures, such as industrial wireless detector platforms, is resource profitability.

10.5.2 Difficulties with storing information and preparation

The accompanying types of input initial processing investigation difficulties are brought on by MIoT data.

- Coordination of data. Different varieties and diverse aspects of knowledge are created in the MIoT. The numerous forms of information must be integrated in order to build effective data statistics methods. However, integrating various forms of MIoT data is rather difficult.
- Reducing redundant work. The MIoT's unprocessed information is characterized by chronological and geographical redundant systems, which frequently lead to content instability and negatively impact the interpretation of that input. It evolves into difficult to reduce redundant information in MIoT information.
- Data reduction as well as data cleansing. In conjunction with the redundancy of information, MIoT information is frequently inaccurate and loud because of broken equipment or instrument problems. The procedure of information cleansing is, nevertheless, made more difficult by the sheer amount of content. That's why it is essential to develop efficient methods for both MIoT data cleaning and compression. Data assessment and significance harvesting heavily rely on the preservation of data. In the MIoT, it might be difficult to create an effective and flexible storage for information infrastructure. The following is a summary of the difficulties with keeping records.
- Consistency and dependability of data retention. Platforms for storing data have to guarantee the legality and longevity of MIoT information. Owing to the enormous volume of facts, it is difficult to meet the aforementioned objectives of big data insights while managing expenses [7].
- Scalability. The flexibility to scale of preservation infrastructure for big data research is a significant problem in addition to memory dependability. The many data kinds, diverse topologies, and size of MIoT's vast data sets make it impossible for traditional databases

to be used for big data insights. To enable large-scale data collection technologies for big data statistics, new storage models must be introduced.

• Efficiency. The reliability of data preservation technologies is yet another issue. Data storage must simultaneously meet the efficiency, dependability, and flexibility criteria to handle the enormous number of parallel requests or inquiries launched throughout the data processing stage.

10.5.3 Data analytics difficulties

Big data analysis for MIoT is particularly difficult because of the enormous quantity, the diverse frameworks, and the extremely large scale. The main difficulties encountered during this era are outlined below.

• The link between time and space in the data. In contrast to traditional data storage facilities, MIoT input is typically connected over both space and time. A new issue arises when trying to handle the input and derive meaningful knowledge from the temporally and geographically connected MIoT signals.
• Dependable data mining techniques. Because of the underlying factors, implementing effective data mining techniques is challenging given the massive amount of MIoT information:
  • Because of the enormous amount of info, it doesn't seem practical to employ traditional multi-pass info-mined algorithms.
  • It is essential for reducing data mistakes and ambiguity brought on by the flawed characteristics of MIoT info.
  • Privacy and security. Regarding privacy and ensuring the security of information throughout the operation of calculations operation is difficult. The vast amount, varied frameworks, and spatiotemporal interconnections of the MIoT media could prevent the use of many typical confidentiality data analysis approaches. To solve the aforementioned problems, new confidentiality information mining strategies must be presented [13].

10.6 EDGE COMPUTING FOR EFFICIENT DATA PROCESSING IN MIOT

Edge computing represents a distributed information technology (IT) infrastructure whereby customer details are handled as near to the starting point as is practical at the infrastructure's edge [14]. In its most basic form, edge processing involves relocating certain memory and processing capacity away from the main data center and towards the actual data stream. As opposed to sending unprocessed information to a centralized information center for the purpose of analysis and processing, that activity is now done wherever the information is really created, either on the working surface within the manufacturing facility, in a storefront, at a large utility, or all throughout an intelligent town. The one and only output of the computer effort at the border that is delivered again to the primary info center for analysis along with additional interactions with people are real-time company perspectives, repair of machinery forecasts, or other useful findings. Thus Figure 10.4, edge technology is changing how businesses and IT employ computers. Examine edge technology in detail, including its definition, operation, the impact of the cloud, applications, compromises, and execution concerns [15].

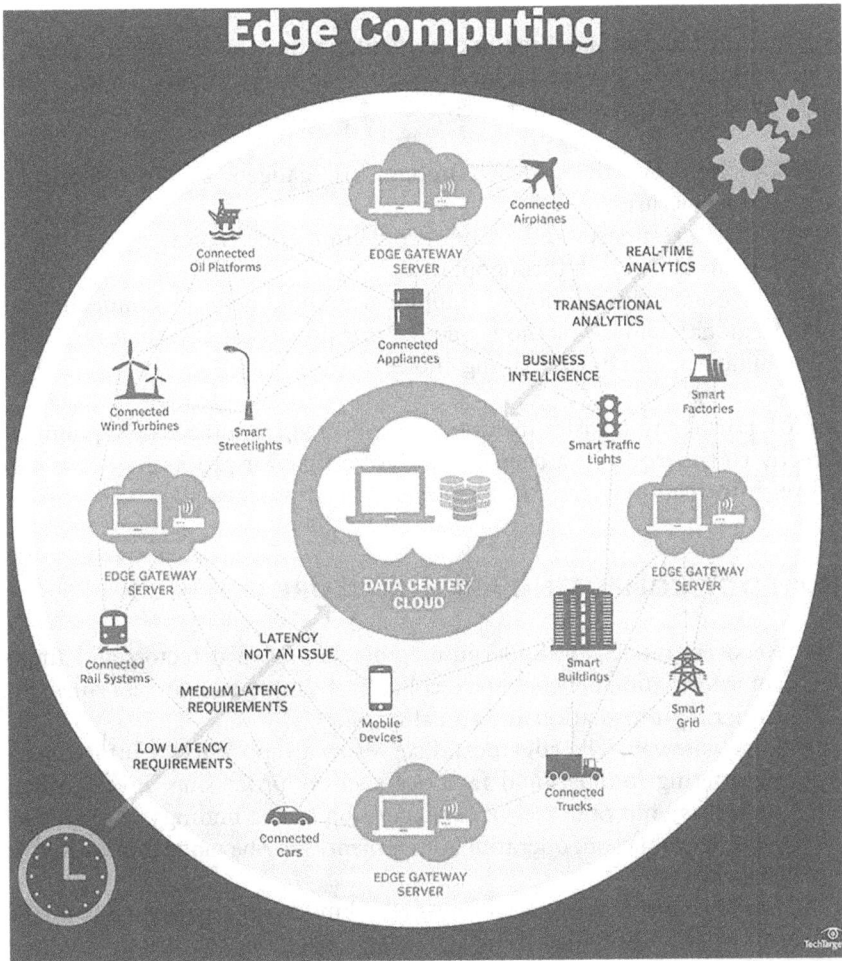

Figure 10.4 Edge computing brings data processing closer to the data source [15].

Figure 10.4, Edge computation, in other terms, is the provision of offerings, computation at the boundary of the network, and information production. To satisfy the pressing demands of the IT sector in terms of adaptable association, real-time enterprise, information optimization, deployment intelligence, safety, and confidentiality, edge technology involves moving the cloud's network, processing, preservation skills, and assets to the edge of the infrastructure. It also provides smart services at the edge to satisfy the necessities of minimal latency and significant bandwidth on the system. Presently, edge computation is a popular topic for study [16–20].

10.7 BENEFITS OF EDGE COMPUTING IN MIOT

For the information handling and archiving requirements of IoT gadgets, edge computation serves as a local origin. The following are some advantages of combining IoT with Edge:

- Lower connectivity delay among IoT gadgets and the main IT infrastructures.
- Quicker responses and greater production effectiveness.
- Increased network throughput.
- The ongoing utilization of equipment without network connectivity.
- Regional analysis methods and ML for processing information, accumulation, and quick making choices.

An IoT portal can transfer information collected from the edge equipment directly to the cloud or a centralized data center for processing, or it can deliver information to the edge devices for local processing.

10.8 EDGE COMPUTING ARCHITECTURE FOR MIOT

- Recovery mode: The underpinning machine's IoT detectors and ambient information are connected collectively, often via wireless rules, for gathering information and transferring.
- Edge gateway: The edge portal serves as a repository of information by offering foundational facilities such as processing, preservation, networks, and other services. According to the findings of the information evaluation, operational judgments may be made thanks to this gathering.
- Cloud center: It is essential for using AI modeling techniques using the information provided by the edge device for big data evaluation and information extraction.

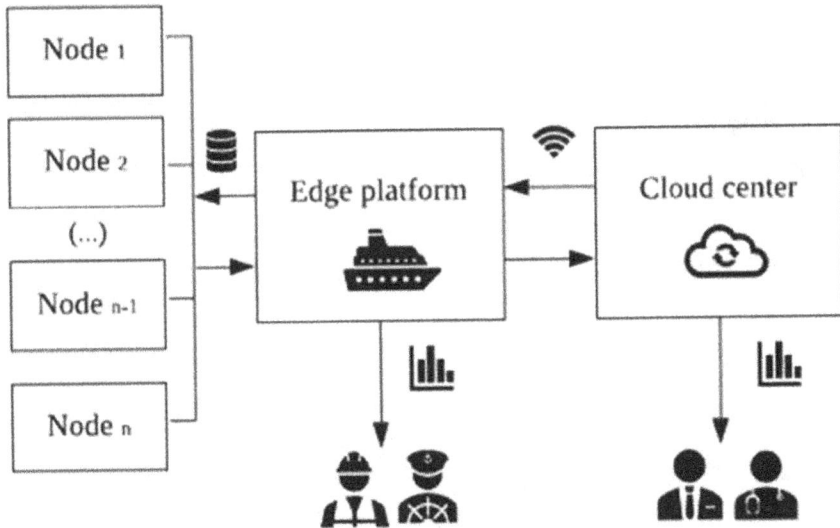

Figure 10.5 An edge computing architecture for IoT data acquisition [21].

Figure 10.5, the suggested edge computation architecture is used to acquire the IoT data collecting. On the edge aspect, the network accomplishes the tasks of analyzing information and storing it. To accommodate more sophisticated demands, it can also be used in conjunction with the cloud center. It is practical in practice to use edge and centralized clouds to take time and size into account in a number of industrial settings [21].

10.8.1 Edge analytics and ML techniques for MIoT

The majority of IoT installations resemble the illustration above. By means of a router, detectors or additional equipment are instantaneously linked to the Internet and deliver unprocessed information to a backend server. On these platforms, ML techniques may be used to forecast a range of situations that would be relevant to management. Whenever excessive devices start to block network congestion, however, things become challenging (Figure 10.6). It's possible that the nearby Wi-Fi is overloaded or that excessive content is being sent to a distant location (and you are unwilling to stump up for it). We can start running simpler machine learning (ML) techniques on a server at home or perhaps on the gadgets themselves to assist in easing some of these problems.

This is referred to as "Edge AI." Instead of using external servers, we're using locally hosted computers or integrated devices to conduct ML methods. Even if the methods used might not be as strong, we could potentially be able to clean up the information before transferring it to a distant place

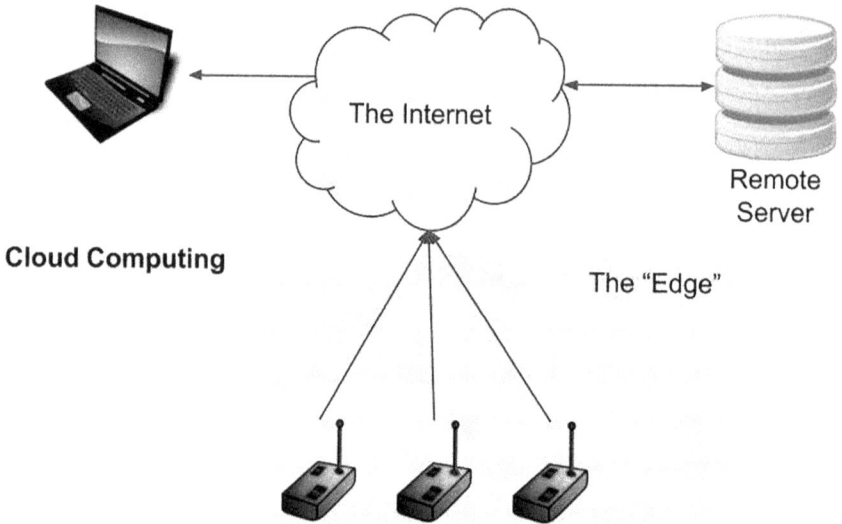

Figure 10.6 Edge without ML [22].

for more research. Smart speakers are a popular illustration of cutting-edge AI. A wake command or phrase (like "Alexa" or "OK Google") has been learned and saved internally on the device that speaks as an ML algorithm. When the wake word or phrase is heard, the intelligent speaker will start "listening." To put it another way, it will start streaming audio data to a distant server so that it can handle the entire request. The consequences of this application of ML are not yet completely understood. But in the years to come, envision witnessing some bizarre and novel computer gadgets. We'll begin to instruct machines on how to acquire knowledge rather than how to accomplish things [22]. Here are a few of the methods:

- Edge Analytics
- Deep Learning
- Reinforcement Learnings
- Transfer Learning
- Federated Learning

10.8.2 Sensor-based data collection

Intelligent sensors are frequently employed to monitor a wide range of various characteristics, making it challenging to begin analyzing the data. You might suddenly find oneself with an incredible quantity of knowledge because devices are now accessible to do anything from measuring a room's climate to notifying you which workstations are taken. However, this knowledge is only truly valuable if one can immediately and readily identify patterns or concerns underlying it.

Throughout a structure, instruments are put in to monitor things like moisture, CO_2 levels, and ownership. Or they might be used for determining things like electricity and the climate by being connected to equipment and wires. They convert live knowledge into electronic facts, that may be supplied to a variety of technology platforms. These programs analyze the facts and produce useful knowledge, enabling you to alter your strategy in response to the newly gathered data. Organizations may get a variety of advantages from using intelligent instruments in this way, such as more effective efficiency, better employee happiness, and the capacity to perform predictive maintenance, averting bigger issues before they arise. Intelligent sensors commonly send information remotely since it is frequently impossible for equipment to be hooked in due to the huge number of devices used in the system and their placement. Although typically sent directly from the device to some type of cloud computing structure for storage and processing, getting the information is not constantly as straightforward. The employment of a terminal, a linking gadget, in every environment, is the most typical method of extracting data from intelligent sensors. Information from the gadgets is received through a gateway, which makes it useful. Wireless data transmission occurs between the instruments and the access point.

The gateway's algorithm then transforms the device's information into JSON and delivers it to the regional or based on the cloud MQTT operator of your choice remotely over Wi-Fi and LTE (4G) or over an Internet connection. When data is saved, analyzed, and shown in the cloud, it can be quickly and easily found and accessed, allowing you to react to any challenges that arise in real time. You can sign up for a variety of IoT cloud channels, such as IBM Watson, Microsoft Azure, and AWS IoT, that can carry out this computation for you. Various gateways, including those offered by Pressac, also include the ability to build a user interface more regionally using a program like Node-Red. As a result, you can create a straightforward interface to show the data that the gateway is receiving without having to create an account on a cloud service [23].

10.8.3 Crowdsourced data collection

Crowdsourcing is the process of gathering data, viewpoints, or creative output from a large number of individuals, typically online. Through the use of crowdsourcing, businesses can access individuals with a variety of skills and viewpoints from around the world while also saving both money and time. Expense reductions, acceleration, and the opportunity to collaborate with individuals who possess abilities that an internal team might lack are all benefits of crowdsourcing. For a major task to be crowdsourced, it often has to be divided into several smaller tasks that a large group of individuals may work on independently [24]. Crowdsourcing is the process of gathering information or assets from a large group of individuals. We may generally divide this into four major categories:

- Wisdom:– The wisdom of crowds is the hypothesis that when it pertains to troubleshooting or determining attributes (such as the mass of an animal or the amount of jelly beans in a jar), big communities of individuals are collectively brighter than individual specialists.
- Creation: A collective attempt to create or construct anything is called crowd creation.
  Instances of this include Wikipedia and other wikis. One other excellent example is open-source technology.
- Voting: By "polling the audience," crowd participation applies the popular vote concept to determine a certain path of motion or strategy.
- Funding: Crowdfunding includes seeking comparatively modest sums from a huge variety of investors in order to raise funds for multiple goals [24].

10.8.4 Mobile data collection

Mobile fact gathering is a technique for gathering both qualitative and quantitative knowledge via a mobile device (such as a phone, tablet, etc.). It has been demonstrated that this strategy improves data collecting efficiency, the efficacy of service shipment, and program staff output. Choice assistance, form reasoning, and checkpoints are examples of characteristics of mobile data collection that enhance the standard of information obtained while also assuring compliance with record gathering and maintenance guidelines. Mobile data collecting should be treated similarly to any other kind of information collection program, despite the fact that it is an efficient platform [25].

10.9 DATA MANAGEMENT TECHNIQUES FOR MIOT

Figure 10.7, Data governance is the method of narrowing down the volume of accessible data to the most crucial knowledge. Huge quantities of knowledge are sent in a variety of formats by numerous gadgets for various uses. Creating and implementing structures, rules, practices, and processes that can address the demands of the whole data lifecycle is necessary to manage all this

IoT content. Smart gadgets operate around the world to streamline processes so we can preserve time. Intelligent objects can gather, transmit, and comprehend knowledge, but sophisticated tools will be needed to assemble the data and identify associations, patterns, and implications.

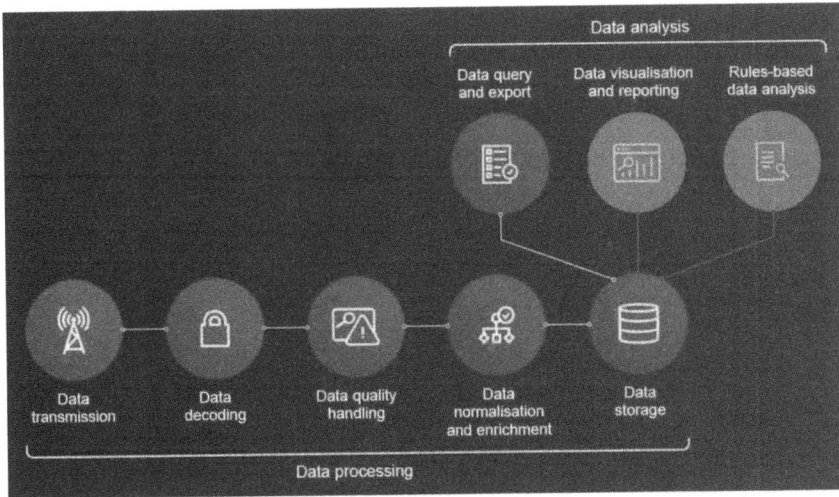

Figure 10.7 Steps in data management [26].

Analysis uses are made on information via connected devices. Dark data refers to knowledge that firms gather and keep but that is mostly static since it is not used for analysis reasons. Consumer demographic statistics, past purchases, customer experience scores, and basic goods details are all included. Dark data is useful to organizations because it enables them to quickly unearth extra insights that will help them properly comprehend their consumers. Field testing is necessary for IoT information handling before an item is released. Field test results are used to refine the layout and provide an item of greater quality. After an item's launch, field data collection aids in the continuing development of the product through software upgrades and the detection of abnormalities. Additionally, this offers crucial insights to aid in developing fresh goods [26].

10.9.1 Data cleaning and integration

Data cleansing, sometimes referred to as data cleaning or scrubbing, is the process of locating and correcting mistakes, duplication, and unnecessary information in a primary collection. Data cleaning is a step in the method of preparing data that produces correct, tenable data that can be used to create trustworthy designs, visualizations, and company choices (Figure 10.8). The method of merging information from multiple places into an individual, cohesive perspective is known as data integration. Integration involves procedures like cleaning, ETL mapping, and modification and starts with the import procedure (Figure 10.9). Analytics technologies can finally create useful, meaningful business knowledge thanks to data integration [28].

Data Cleansing

| DEDUPE YOUR DATA | REMOVE IRRELEVANT OBSERVATIONS | MANAGE INCOMPLETE DATA | IDENTIFY OUTLIERS | FIX STRUCTURAL ERRORS | VALIDATE YOUR WORK |

Figure 10.8 Data cleansing step [27].

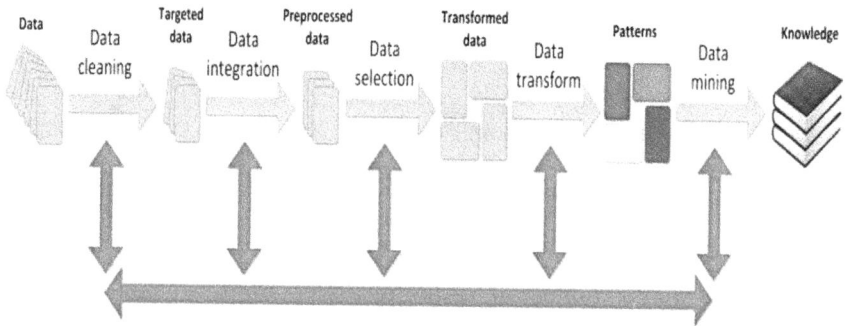

Figure 10.9 Steps in processes of knowledge discovery [29].

10.9.2 Data storage and retrieval

To meet customer data demands, content preservation and recovery (IS&R) deals with the representation, preservation, organization, and accessibility of material objects. Archives and extraction, in other words, is the act of looking for pertinent records within a big, unorganized corpus in order to meet the user's demands for knowledge. Additionally, in Figures 10.10 and 10.11, it is a tool that identifies and chooses a particular category of objects from a group of objects to suit the consumer's needs [30].

10.10 REAL-TIME PROCESSING AND ANALYSIS TECHNIQUES FOR MIOT

A real-time device or technology receives the data you provide, interprets it, and produces a useful result in a matter of a few seconds. Real-time examining involves handling information that comes from numerous resource streams continuously and with extremely little lag. The necessity of analyzing this fact rapidly in order to gain a comparative edge is transforming into important as businesses gather increasing quantities of information.

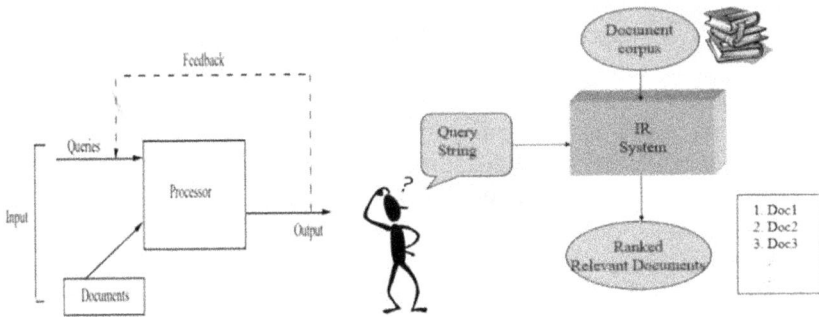

Figure 10.10 A typical information retrieval system [31].

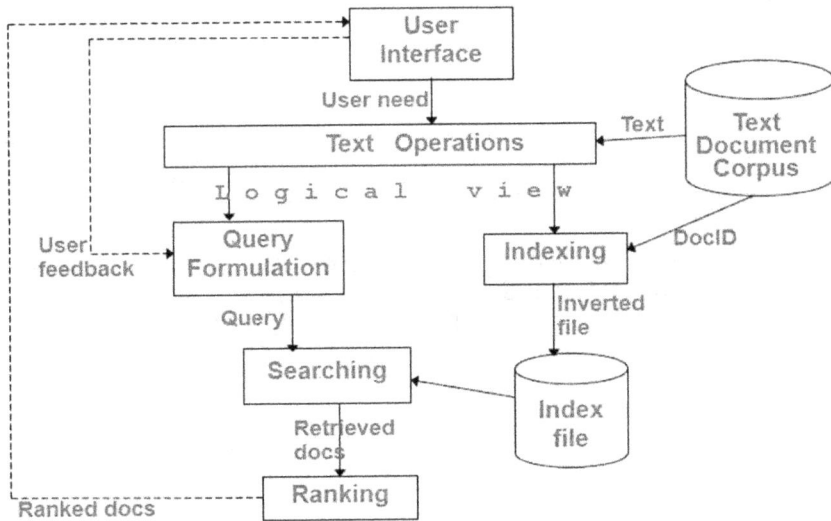

Figure 10.11 Retrieval process [32].

Processing input in real-time is necessary for many businesses, including network tracking, information security, banks, and e-commerce, in order to identify scams, track daily activities, and identify prospective business possibilities [33].

The subsequent four elements make up the framework for real-time information handling:

- Real-time Message Ingestion:
- Stream Processing
- Analytical Data Store
- Analysis and Reporting.

There are certain similarities between the live information broadcasting devices and systems that have recently joined the competition [34].

- In-memory processing: To handle streams, RAM is essential. In the case of enormously concurrent systems, when the abilities of numerous machines are combined, disk-based solutions literally cannot handle flows in real time. In-memory computing systems have become the de facto norm for creating programs that handle real-time data broadcasting because of the falling cost of RAM.
- Open source: The foundation for numerous current information broadcasting devices and methods was laid by the freely available sector. Real-time flow treatment solution suppliers frequently develop functions and parts to make them organization-ready, but beneath there is frequently a free-standing aspect.
- Cloud: Numerous cloud-based services produce information for real-time distribution handling.
- Window-based processing: Bigger data is processed by breaking it up into fewer components. Just like Figure 10.12 shows the event-based processing.
- Event-based Processing:
- Event source: Notifications are sent to a particular event handling mechanism by an event producer. Simple RFID detectors and motors, business processes, CICS programs, and CICS system parts are a few examples of event inputs [35].

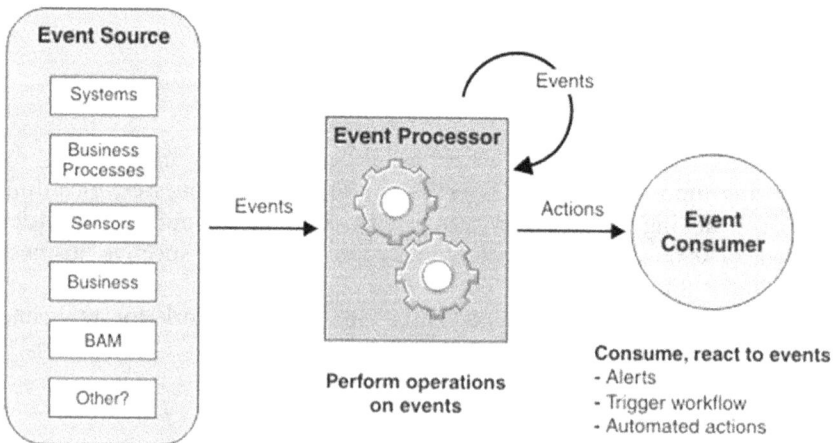

Figure 10.12 Event processing architecture.

- Event processor: The incident handling system can take the following actions with respect to incidents:
  - Enhance a single activity in an easy way.
  - Include details regarding the cause of the incident.
  - Compare event characteristics to numerous individual occurrences from various event providers to create an entirely novel, evolved, or compound occurrence.

The corresponding client can then access the edited occurrence.

- Event consumer: The occurrence's customer responds to it. A client of an occurrence might be as basic as refreshing an organization's display or file system, or as complicated as necessary by doing new administrative operations in response to the occurrence [35].
- Time-based processing: Time-based algorithms are employed to alter and reconstruct the environment. The delay. RAM storage of recorded sound produces electronic lag. The sound is listened to from storage after a predetermined period.

10.11 PREDICTIVE ANALYTICS FOR MIOT

Forecasting and Resource Optimization A subset of modern analysis called forecasting uses past information along with mathematical modeling, data extraction, and ML to forecast upcoming occurrences. By employing predictive statistical analysis, businesses may recognize hazards as well as possibilities by looking for variations in this data [36].

10.11.1 Time series prediction and analysis for MIoT

Making scientific projections with reference to evidence with prior moment stamps is known as time series projections. It entails creating algorithms via research into the past and utilizing them to draw conclusions and guide tactical decisions in the coming years. The fact that the upcoming result is wholly unknown at the beginning of the task and can only be anticipated via extensive research and grounded in evidence is a crucial difference in predicting [37].

10.11.2 Forecasting analytics for capacity optimization

The method of assigning and controlling assets in a highly effective manner is known as asset optimization. Assets can be physical (workers) and technological (software, equipment) in various organizations, including IT corporations and digital agencies [38]. An optimization issue must be solved using a goal, factors, and limitations. The objective is to determine the best value for each factor, which might have a variety of possibilities.

10.12 ML FOR MIOT: ANOMALY DETECTION AND PREDICTIVE MAINTENANCE

IoT and ML unlock ideas from input that would normally be concealed, enabling quick, automatic reactions, and enhanced making choices. By consuming photos, footage, and recordings, ML for IoT may be used to forecast upcoming patterns, identify abnormalities, and boost expertise. By utilizing complicated techniques to analyze vast amounts of information, ML may assist in demystifying the unconscious trends in IoT data. In crucial operations, programmed devices applying statistically determined responses can augment or completely replace manual procedures [38].

10.12.1 Anomaly detection using ML

Identifying those uncommon things, facts, occurrences, or findings that raise concerns by standing out from the remainder of the information pieces or views is the method of recognizing anomalies. Outlier identification is another name for identifying anomalies. Two categories may be used to broadly classify anomaly detection strategies:

- Supervised Anomaly detection
- Unsupervised Anomaly detection o Supervised Anomaly Detection

The annotated data with labels, which includes both normal and anomalous information, is required for supervised identification of anomalies in order to build a prediction classifier. Neural networks, support vector machines, k-nearest neighbors, Bayesian networks, decision trees, etc. are a few of the popular supervised approaches.

- Unsupervised Anomaly Detection
  Unsupervised: There is no need for labeled learning information for anomaly recognition. These methods are predicated on the following two notions:
  - The majority of network links come from regular congestion, and there is very little unusual content.
  - Statistics show that malicious traffic differs from regular transmission.

10.12.2 Predictive maintenance using ML

An idea called predictive maintenance (PdM) is used to administer maintenance schedules for facilities by foreseeing their breakdowns using data-driven methods. In these cases, information is gathered over time to track the gadget's condition [39]. One of two methods [40] can be employed to deal with a sustainability issue according to the information acquired by data investigators [40]:

- Supervised learning if the organization's information set contains labeled inability incidents.
- Unsupervised learning when the information set lacks labeled loss occurrences.

10.12.3 Security and privacy challenges in MIoT data analytics

- Enormous needs for data creation and preservation – Since the assembled data produced by thousands of IoT instruments is typically enormous, managing and storing such details can be quite costly.
- The difficulty of information – Unique kinds, forms, and measurements of information from numerous IoT gadgets make it complicated and challenging to handle and clean up.
- Protection – Organizations must implement several safety precautions to guard recorded IoT structure information from hackers' hacking and breaches, particularly those that involve consumer information.
- False information – False readings from malfunctioning IoT gadgets complicate the processing of this kind of information. The findings provided by IoT monitoring solutions become problematic when there are a high number of defective IoT gadgets.
- Assembling a skilled analysis workforce – To carry out analyzing procedures on IoT information and gain useful knowledge, organizations must employ data engineers and researchers [41].

10.13 OPEN RESEARCH ISSUES

Table 10.1 Open issues, challenges, and future directions of 6G-enabled IoT [3,42]

6G-enabled IoT open issue	6G-enabled IoT challenges	6G-enabled IoT future directions
Transportation	• Storage of information and relocation • Safety of data, accessibility, and asset	• The flexibility of IoT gadgets • The potential to fully mobilize and install huge edge terminals in the 6G infrastructure.

(Continued)

Table 10.1 (Continued) Open issues, challenges, and future directions of 6G-enabled IoT [3,42]

6G-enabled IoT open issue	6G-enabled IoT challenges	6G-enabled IoT future directions
methods for unpredictable accessing	• Information delivery that is trusted - Data transport that uses fewer resources • Data transport with high spectral efficiency	• Minimize wireless resource expenses for interaction. • Existing random-access methods and standards have the potential to work together to create more efficient wireless networks.
Use of energy and power	• The influence of every component in an individual subarray on the interpersonal connection. • Managing resources • Managing temperature • Erratic subarray tile calibrating	• Utilizing intelligent 6G connections for transforming electricity in manufacturing applications; watching the most effective impact of power strategy when the highest possible energy expenses are constrained; investigating the way the deployment of multi-antenna structures impacts the intake of power.
Protection (which includes gain access, confidentiality, etc.)	• Defects in networks • Data communication, data unloading, facts safety, and network assaults • Permission and verification	• Utilizing ML to find network weaknesses
Synthetic intelligence	• The difficulty of implementing computer and AI skills accessibility	• Improving rebroadcast methods with AI. • Cloud and edge computing for smart cooperation
Trustworthy interactions	• Extreme dependability • Exceptional dependability • Verified spectrum pooling	• To increase efficacy and dependability, upcoming investigation should use multiple forms of data on shipping, such as surveillance footage, radar detectors, and BeiDou geolocation. • A fair and trustworthy radio transmission system that is operating system-defined • Dependable

(Continued)

Table 10.1 (Continued) Open issues, challenges, and future directions of 6G-enabled IoT [3,42]

6G-enabled IoT open issue	6G-enabled IoT challenges	6G-enabled IoT future directions
		• And low-latency interactions may be provided by using memory and processing.
Technical difficulty	• Customer engagement • Expandable framework • Movable foundation	• Develop a sequential interference cancellation (SIC) approach that is complicated enough. • Developing a low complexity identification method with suitable restrictions to calculate the likelihood that each consumer's subcarriers in the channel-block will be activated. • The use of flexible conflict analysis enhances • 6G users' quality of experience independent of the needs of the applications.

10.14 CONCLUSION

The total number of demands for various connections, the login achievement and collapse pace of responsive and unresponsive ports depending on various limitations variables, the message dissemination importance of each node, and the spreading delay time are investigated by analyzing the enormous accessibility of mobile data IoT gadgets in 6G innovation. Furthermore, the information dissemination and analysis stage determined by the use of 6G is brought out by contrasting the delivery power use and access duration whenever there is a regional information cache and when cloud seeking is necessary. Yet, this method does not take into account how disturbances will affect the information. The influence on information distortion will be examined in the near future to improve the accuracy and efficiency of IoT data processing. As a result, 6G-based big data research can offer top-notch technological assistance for transferring information among huge IoT gadgets and offer suggestions for how to use 6G innovation in massive IoT. It is also very important for the growth of the IoT in general.

REFERENCES

[1] Kranz, G., & Christensen, G. (2022). What is 6G? Overview of 6G networks & technology. Networking. https://www.techtarget.com/searchnetworking/definition/6G#:~:text=6G%20(sixth%2Dgener ation%20wireless),support%20one%20microsecond%20latency%20communications.

[2] Massive Io T: Tech overview, business opportunities and examples. (n.d.). Thales Group. https://www.thalesgroup.com/en/markets/digital-identity-and-security/mobile/massiveiot#:~:text=Massive%20IoT%20refers%20to%20the,new%20ways%20to%20solve%20probl ems.

[3] Memon, M. A., Soomro, S., Jumani, A. K., & Kartio, M. A. (2017). "Big data analytics and its applications." arXiv preprint arXiv: 1710.04135.

[4] Guo, Fengxian, Richard Yu, F., Zhang, Heli, Li, Xi, Ji, Hong, & Leung, Victor C. M. (2021). "Enabling massive IoT toward 6G: A comprehensive survey." IEEE Internet of Things Journal 8(15): 11891–11915. https://www.csit.carleton.ca/~fyu/Papers/IoT_6G.pdf.

[5] MindMajix Technologies. (2023, April 3). Big data in IoT | Role of big data in IoT 2023. Mindmajix. https://mindmajix.com/big-data-in-iot.

[6] Chai, W., Labbe, M., & Stedman, C. (2021). Big data analytics. Business Analytics. https://www.techtarget.com/searchbusinessanalytics/definition/big-data-analytics#:~:text=Why%20is%20big%20data%20analytics,personalization%20and%20impr oved%20operational%20efficiency.

[7] Sharma, S. K., & Wang, X. (2017). "Live data analytics with collaborative edge and cloud processing in wireless IoT networks." IEEE Access 5: 4621–4635, doi: 10.1109/ACCESS.2017.2682640.

[8] Wahla, Arfan, Chen, Lan, Wang, Yali, Chen, Rong, & Wu, Fan. (2019). "Automatic wireless signal classification in multimedia internet of things: An adaptive boosting enabled approach." IEEE Access 7: 160334–160344, doi: 10.1109/ACCESS.2019.2950989.

[9] Altman, Naomi S. (1992). "An introduction to kernel and nearest-neighbor nonparametric regression." The American Statistician 46 (3): 175–185.

[10] Alvaro, Peter, Condie, Tyson, Conway, Neil, Elmeleegy, Khaled, Hellerstein, Joseph M., & Sears, Russell. (2010). "Boom Analytics: Exploring Data-centric, Declarative Programming for the Cloud." In Proceedings of the 5th European Conference on Computer Systems (EuroSys), pp. 223–236. doi: 10.1145/1755913.1755937.

[11] Apache. (2014). "Hadoop MapReduce." https://hadoop.apache.org/docs/r2.6.0/hadoop-mapreduce-client/hadoop-mapreduce-client-core/MapReduceTutorial.html

[12] Azadeh, A., Saberi, M., Kazem, Ebrahim Ipour, V., Nour Mohammad Zadeh, A., & Saberi, Z. (2013). "A flexible algorithm for fault diagnosis in a centrifugal pump with corrupted data and noise based on ANN and support vector machine with hyper-parameters optimization." Applied Soft Computing 13(3): 1478–1485.

[13] Dai, Hong-Ning, Wang, Hao, Xu, Guangquan, Wan, Jiafu, & Imran, Muhammad. (2020). "Big data analytics for manufacturing internet of things: Opportunities, challenges and enabling technologies." Enterprise Information Systems 14(9–10): 1279–1303, DOI: 10.1080/17517575.2019.1633689.

[14] What is IoT edge computing? (n.d.). https://www.redhat.com/en/topics/edge-computing/iot-edge-computing-need-to-work-together#how-are-iot-an d-edge-related.

[15] Bigelow, Stephen J. (n.d.). What is edge computing? Everything you need to know. Techtarget. https://www.techtarget.com/searchdatacenter/definition/edge-computing.

[16] "The Internet of Things reference model" in Internet of Things World Forum, Chicago, IL: Cisco Inc, pp. 1–12, Oct. 2014.

[17] Sun, X., & Ansari, N. (2016). "Edge IoT: Mobile edge computing for the internet of things." *IEEE Communications Magazine* 54(12): pp. 22–29.

[18] Alrawais, A., Alhothaily, A., Hu, C., & Cheng, X. (2017). "Fog computing for the internet of things: Security and privacy issues." *IEEE Internet Computing* 21(2): 34–42.

[19] Kang, J., Yu, R., Huang, X., & Zhang, Y. (2018). "Privacy-preserved pseudonym scheme for fog computing supported internet of vehicles." *IEEE Transactions on Intelligent Transportation Systems* 19(8): 2627–2637.

[20] Mouradian, C., Naboulsi, D., Yangui, S., Glitho, R. H., Morrow, M. J., & Polakos, P. A. (2018). "A comprehensive survey on fog computing: State-of-the-art and research challenges." *IEEE Communications Surveys and Tutorials* 20(1): 416–464.

[21] Plaza, Marta, Gil, Ana, Rodríguez, Sara, Prieto-Tejedor, Javier, & Corchado Rodríguez, Juan. (2021). *Integration of IoT Technologies in the Maritime Industry*. Springer. doi: 10.1007/978-3-03053829-3_10.

[22] ShawnHymel. (2022). What is edge AI? Machine learning + IoT. Digi-Key Electronics. https://www.digikey.com/en/maker/projects/what-is-edge-ai-machine-learningiot/4f655838138941138aaad62c170827af.

[23] Woodward, E. (2021, November 12). How to manage data collected from smart sensors. Pressac Communications. https://www.pressac.com/insights/how-to-manage-data-collected-from-smart-sensors/#:~:text=The%20more%20common%20way%20of,sensors%20to%20the%20gatewa%20y%20wirelessly.

[24] Hargrave, M. (2022). Crowdsourcing: Definition, how it works, types, and examples. Investopedia. https://www.investopedia.com/terms/c/crowdsourcing.asp#tocwhat-is-crowdsourcing.

[25] Mobile data collection | What is mobile data collection? (n.d.). Dimagi. https://www.dimagi.com/mobile-data-collection/#:~:text=Mobile%20data%20collection%20is%20a,effectiveness%2C%20and%20 program%20staff%20 performance.

[26] Ks, R. (2021, February 10). Data management systems for the IoT devices | IoT device management. Profit from IoT | IoT India. https://iot.electronicsforu.com/content/techtrends/data-management-systems-iot-devices/.

[27] Data cleaning in data mining. (n.d.). Javatpoint. https://www.javatpoint.com/data-cleaning-in-data-mining#:~:text=Usage%20of%20Data%20Cleaning%20in%20Data%20Mining&text=Data%20Integration%20is%20the%20process,moving%20to%20the%20final%20destination.

[28] Talend. (n.d.). What is data integration? Talend. https://www.talend.com/resources/what-is-data-integration/#:~:text=Data%20integration%20is%20the%20process,%2C%20ETL%20mappin g%2C%20and%20 transformation.

[29] Hot Dervic, Elma, Vesna, Popović-Bugarin, Topalovic, Ana, & Knežević, Mirko. (2016). Generating thematic pedologic maps by using data mining and interpolations. In *3rd International Conference on Electrical, Electronic and Computing Engineering*, pp. 1–7.

[30] 1.2. What is information storage and retrieval (ISR)? (n.d.). Wachemo University e-Learning Platform. https://wcu.edu.et/.

[31] 1.4. IR System Architecture. (n.d.). Wachemo University e-Learning Platform. https://wcu.edu.et/

[32] 1.7. The Retrieval Process (Basic Structure of an IR System). (n.d.). Wachemo University eLearning Platform. https://wcu.edu.et/.

[33] Sharma, D. (2022, December 29). What is real-time processing? A quick-start guide. Hevo. https://hevodata.com/learn/real-time-processing/#i2.

[34] Hazelcast. (2019, September 25). What is real-time stream processing? Hazelcast. https://hazelcast.com/glossary/real-time-stream-processing/#:~:tex t=Real%2Dtime%20stream%20processing%20is,data%20is%20generated %20or%20published.

[35] IBM Documentation. (n.d.). https://www.ibm.com/docs/en/developer-for-zos/ 9.1.1?topic=concepts-what-is-event-processing.

[36] What is predictive analytics? (n.d.). IBM. https://www.ibm.com/topics/predic tiveanalytics#:~:text=Predictive%20analytics%20is%20a%20branch,to%20 identify%20risks%2 0and%20opportunities.

[37] Time series forecasting: Definition, applications, and examples. (n.d.). Tableau. https://www.tableau.com/learn/articles/time-series-forecasting#:~:tex t=Time%20series%20forecasting%20occurs%20when,drive%20future%20 strategic%20decision%2Dmaking.

[38] How can machine learning (ML) optimize the IoT? (n.d.). Software AG. https://www.softwareag.com/en_corporate/resources/iot/article/machinelearn- ing.html#:~:text=behavior%20and%20events.-,What%20is%20machine%20 learning%3 F,ingesting%20image%2C%20video%20and%20audio.

[39] Kane, Archit P., Kore, Ashutosh S., Khandale, Advait N., Nigade, Sarish S., & Joshi, Pranjali P. (2022). "Predictive maintenance using machine learning." arXiv preprint arXiv:2205.09402.

[40] Barinov, A. (2022). Using a traditional machine learning approach for predictive maintenance. *The AI Journal*. https://aijourn.com/ using-a-traditional-machine-learning-approach-for-predictive-maintenance/.

[41] WINIX Technologies. (2018, June 28). 6 main challenges facing IoT. Medium. https://medium.com/@winix/6-main-challenges-facing-iot-b6055bdf6782.

[42] Hosseinzadeh, Mehdi, Hemmati, Atefeh, & Rahmani, Amir Masoud. (2022). "6G-enabled internet of things: Vision, techniques, and open issues." *CMES-Computer Modeling in Engineering & Sciences* 133(3): 509–556.

Chapter 11

Healthcare IoT networks using LPWAN

Maysa Khalil and Qasem Abu Al-Haija
Princess Sumaya University for Technology

Samir Ahmad
MTN Consulting LLC.

11.1 INTRODUCTION

The merging of different technologies enriches human life with excellent value solutions for many problems and improves the existing systems or applications to enhance other issues.

The growth of a smart environment with intelligent health services results from the rapid development of sensor technologies and rising computing capacity with effective learning techniques in the Internet of Things (IoT) [1].

The migration of people from rural to urban areas significantly impacts healthcare services because urban cities are working to provide a better and healthier environment using digital transformation and digital technologies [2].

The foundation of new healthcare applications recognizes human activity, which can be done with wearable or ambient sensors to collect data on human activity and health [3].

IoT technology has made a great revolution in the health sector.

This chapter explains the details of merging the IoT and LPWAN technologies to benefit the health sector the most.

11.2 IoT

IoT, which stands for Internet of Things, is the extension and growth of the Internet, connecting various devices to enable intelligent identification, positioning, tracing, monitoring, and network management through the exchange of information [4].

In 2005 the International Telecommunication Union (ITU) formally announced the concept of the IoT; it was defined as an extensive, intelligent network created by merging the Internet with a range of information sensing devices, including radio frequency identification devices, a variety of sensor nodes, and a variety of wireless communication devices [5].

DOI: 10.1201/9781003426974-11

203

11.2.1 Stages of IoT

There are three stages of IoT, it is clarified in Figure 11.1, and they are:

1. **Local-area IoT:** Things are connected in near-field communication like Bluetooth, RFID, and NFC, which only permits essential information exchange among a small number of non-intelligent devices in the local area.
2. **Wide-area IoT:** At this stage, data transmission transcends geographical boundaries to enable widespread IoT coverage and strong connections among somewhat advanced intelligent devices. This is achieved using communication technologies such as 2G, 3G, 4G, and 5G. Like: agricultural climate monitoring and smart meters.
3. **Global intelligent IoT:** This stage is still in the way; it enables the injection of intelligent genes into the Internet of Things from all aspects of perception, analysis, and application. It is used in smart factories and autonomous driving [6].

11.2.2 Applications of IoT

Based on the above stages, IoT has many applications; they are applied widely in many fields [7], mainly in the following fields:

- Creating better enterprise solutions
- Integrating smarter homes
- Innovating agriculture
- Building smarter cities

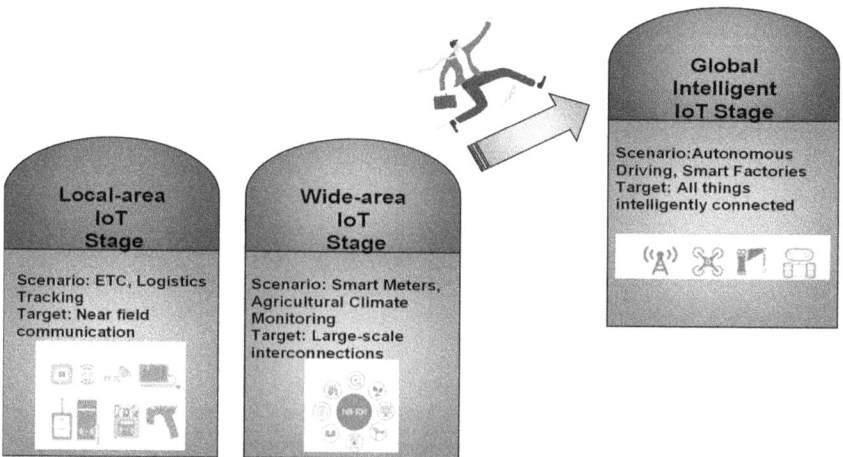

Figure 11.1 Three stages of IoT.

- Upgrading supply chain management
- Transforming healthcare
- Installing smart grids
- Revolutionizing wearables
- Integrating connected factories
- Reshaping hospitality.

Still, the field which gets a high benefit from these applications is the health sector. This sector needs fast speed, low power, and high-quality mentoring and sensors to follow the patient's status, take care of their situation and take suitable action/decision based on their status. Human life and ability are the most important goals, so the IoT is a great innovation to achieve such a goal.

However, IoT applications' success depends on combining IoT and other technologies like LPWAN.

Wearables used to monitor patients with chronic diseases like diabetes, heart disease, or asthma are one example of a healthcare IoT network utilizing LPWAN technology.

Figure 11.2 shows different types of wearable devices.

IoT architectures cannot effectively operate in locations with weak or unstable Internet. At the same time, LPWAN technologies can overcome the inadequate network infrastructure because of its long-range communication protocols [8].

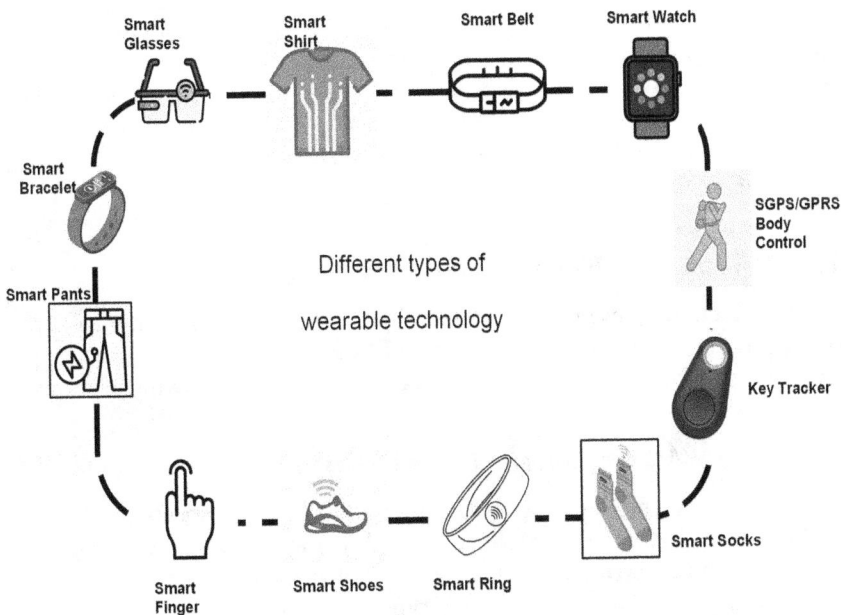

Figure 11.2 Different types of wearables.

LPWAN technology makes it possible to gather real-time patient vital sign data like heart rate, blood pressure, and oxygen levels, which can then be securely communicated to healthcare professionals for analysis and treatment suggestions. Facilitating early detection of health concerns and lowering the likelihood of consequences can enhance patient outcomes [9].

Another advantage of LPWAN technology is that it enables large-scale IoT installations, which makes it perfect for healthcare systems that require communication over vast distances and cover several locations. So, what is the LPWAN, and how does it work?

11.3 LPWAN

LPWAN stands for Low Power Wide Area Network, a type of wireless network technology designed to support low-bandwidth, long-range communication between devices and high signal penetration through buildings or other obstacles. LPWAN is specifically designed for IoT devices that require long battery life and operate in remote locations or areas with poor cellular coverage, so it has extended coverage. Indeed, it offers more benefits such as connecting several devices, designed to send low data rates, energy-efficient hardware design, and affordable equipment prices [10].

11.3.1 LPWAN architecture

LPWAN architecture contains IoT devices, gateways, network servers, and application servers. These components are wired and wireless access between connection and components to the Internet and the cloud for data backup and processing [11].

Figure 11.3 shows the basic architecture of LPWAN and how it connects to the IoT devices.

11.3.2 Types of LPWAN technologies

LPWAN technologies are classified into two categories based on the operating spectrum, whether licensed or not [12]. Figure 11.4 shows these types based on this classification; they are elaborated in the following sections.

11.3.2.1 LPWAN technologies in the Licensed Spectrum

Two LPWAN technologies are working with a licensed spectrum: LTE-M and NB-IoT; they are expensive compared to the LPWAN technologies in the unlicensed spectrum because of the cost of spectrum license. Figure 11.4 summarizes the LPWAN technology types.

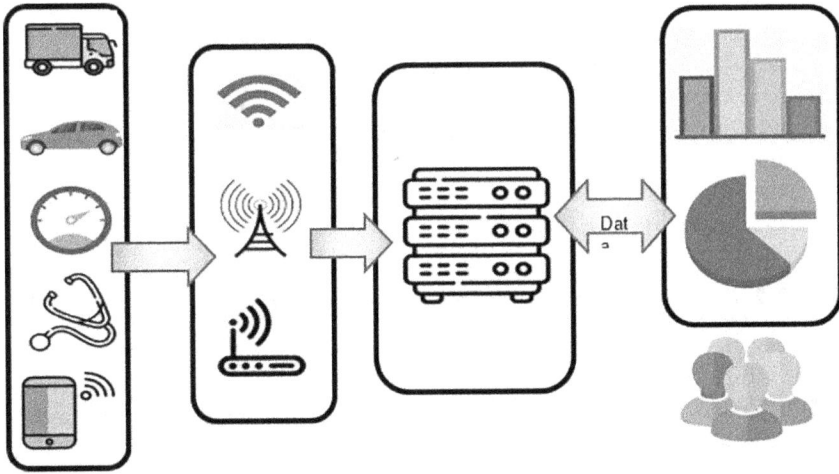

Figure 11.3 LPWAN architecture.

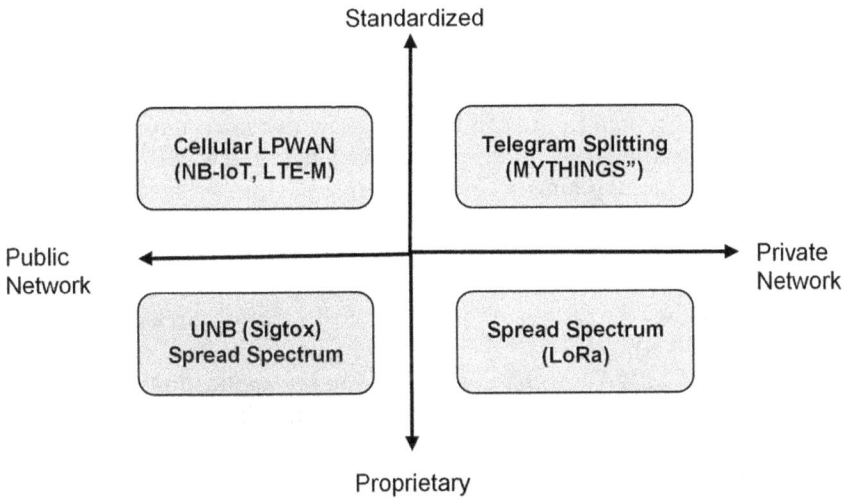

Figure 11.4 LPWAN technologies types.

- **LTE-M** stands for Long-Term Evolution for supporting machine-type communication (MTC); it provides data rates up to 1 Mbps. It is compatible with 3G and LTE.
- LTE-M technology upgrades the legacy LTE to make the technology suitable for IoT applications in terms of functionality and performance.

- **NB-IoT** stands for Narrow Band IoT: Its spectrum (700–900 MHz) in the cellular infrastructure provides data rates up to 100 Kbps. It uses FDMA (Frequency Division Multiple Access) in the uplink and OFDMA (Orthogonal FDMA) in the downlink.
- Wider signal coverage, increased flexibility, reduced energy usage, increased simplicity, and cost efficiency are all goals of this technology. It is compatible with the wireless communication systems GSM, GPRS, and LTE but not with 3G [8].

11.3.2.2 LPWAN technologies in unlicensed spectrum

Another two LPWAN technologies are working with an unlicensed spectrum, which Ultra Narrowband (UNB) and Long Range (LoRa).

Due to the usage of unlicensed frequencies, the network of these technologies is accessible to users without the radio frequency regulators' approval.

- **LoRa** is exceptional because it can operate on public or private networks and adapt to various business models. Offers security in written protocols, an open standard with cheap ownership costs, and interoperability.
- Implementing a LoRa network is quite straightforward, and this technology benefits consumers with minimal setup and maintenance costs.
- It can transmit unlimited daily messages with a maximum data rate of 50 kbps to thousands of devices from a single gateway.
- **UNB** technology aims to offer widespread connection for extensive IoT networks with a low power advantage [13]. UNB, known as Sigfox, referred to the name of the company that created it.

Figure 11.5 shows the system architecture of this Sigfox wireless body area networks (WBAN).

Each type of LPWAN technology is used in the application based on its properties and features and how it will be useful. Table 11.1 compares the four LPWAN technologies in the essential features [14].

11.3.3 Applications of LPWAN

A wide range of innovative and intelligent applications, including environment monitoring, smart cities, smart utilities, agriculture, health care, industrial automation, asset tracking, logistics, and transportation, can be implemented using the promising LPWAN technology because the key advantage of it is its low power consumption, so the devices operate on a single battery charge for years; this makes them perfect for applications that need low maintenance and long-term operation achieved using a combination of energy-efficient hardware design with low data rates and long-range communication.

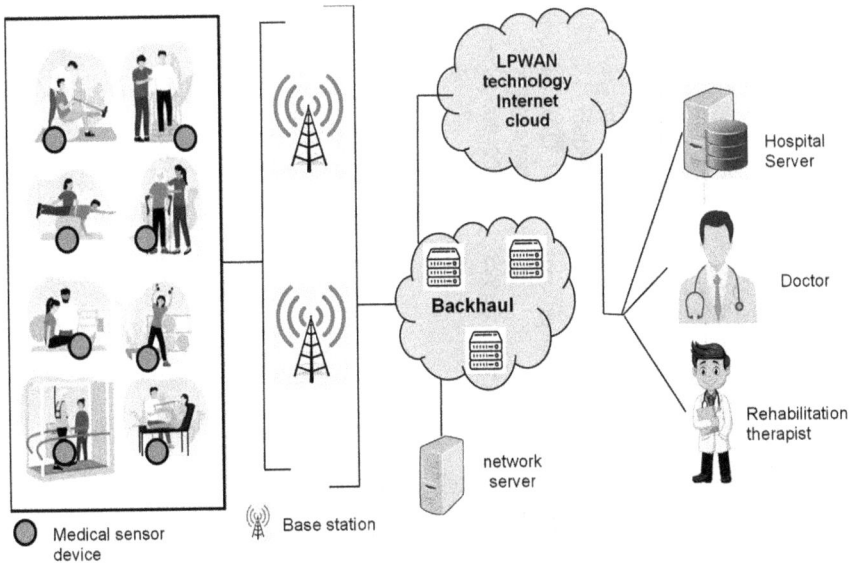

Figure 11.5 Sigfox wireless body area networks (WBAN).

Different applications have varying needs. Of course, the main motivators for all LPWAN applications are coverage, capacity, cost, and low-power operation. Any LPWAN solution, however, may necessitate important trade-offs between various criteria, such as coverage against cost [15].

Also, while specific applications, like meters, are relatively uniform, others contain a vast array of heterogeneous devices with diverse requirements. However, specific applications need additional features, such as the ability to collaborate with other technologies and support for speech, among others. As a result, one LPWAN solution may be tailored to a small number of applications, while another may be created to address a wide variety of applications and qualities [16].

11.4 HEALTHCARE IoT NETWORKS USING LPWAN

To use LPWAN in the health sector, some requirements should be satisfied before the implementation of healthcare IoT networks using LPWAN [17]. These requirements are:

11.4.1 Handling different traffic classes

The network should be able to prioritize the traffic, as there are two types of traffic, which are:

Table 11.1 LPWAN technologies comparison

Property	NB-IoT	LoRaWAN	UNB	LTE-M
Power consumption	11 years	13 years	13 years	Ten years
Range	15 km	<11	<13 km	<11 km
Transmitted power	23 dBm	14 dBm	14–27 dBm	23 dBm
Physical Layer	QPSK Sectorized one antenna	CSS Omnidirectional one antenna	BPSK+UNB Omnidirectional one antenna	16 QAM Sectorized antenna
Spectrum [MHz]	700–900 3 Depl. Modes	Europe: 868,433 US: 915 Asia: 430	Europe: 868 US: 902 Asia: 923	700–900 In-band
Bandwidth [kHz]	3GPP200 standalone 180 in-band	US 500-250-125 Europa 250-125	192 kHz	1.4 MHz
Data rate [bps]	67k UL30k DL	0.3–50k ADRDifferent for DL	100 or 600	< 1 Mbps
QoS	Very High	High	High	Very High
Latency	medium	medium	medium	low
Mobility	No	Yes Non-GPS	Yes Non-GPS	Full GPS-based
module price	10–15 $	9–12 $	5–10 $	7–12 $
connectivity	1 $/ month/100 kB	free in public 500 $ construction	< 1 $ / month	3–5 $/mon /1 MB

- Urgent traffic: It is the traffic for emergency cases
- Traffic that could be delayed: like remote monitoring of regular patients

11.4.2 Scalability

The ability to expand the network and increase the number of IoT devices when needed.

11.4.3 Coverage

The ability to cover all the planned areas to serve all the patients in this area; any inability in this coverage could affect the patient's health negatively. The coverage area should be up to 40 km in the rural area and up to 5 km in the urban area.

11.4.4 Efficient energy

Providing long battery life is one of the essential requirements in the health sector.

11.4.5 Software and device simplicity

A healthcare system would benefit from LPWAN devices with minimal operating costs and extensive coverage. Also, the software should only generally demand a little processing power because this would raise the device's power consumption. This would make network deployment easier and aid in lowering network operating expenses.

11.4.6 Low cost

LPWAN cost should be low in deployment and operation to be effective, and both the patient and the service provider would benefit from this.

11.4.7 Privacy and security

This is too important; the requirements for authorization, authentication, data security, and confidentiality must be supported to prevent malicious users from abusing the healthcare system. These security services should prevent major Denial of Service (DoS) attacks against these devices [18].

11.4.8 Patient location

LPWAN should be connected to a Global Positioning System (GPS) or use an intelligent localization method to identify the patient's location to provide critical care when needed [19].

If these requirements are satisfied, a wide range of applications will be created by combining IoT and LPWAN. These applications monitor the person's vital signs or any signs needing health provider care and send messages to notify the health providers without interfering with the person's normal daily activities. Figure 11.6 shows some of these applications.

The monitoring systems [20] are a good example of these applications as they monitor body temperature, electrocardiogram, oxygen saturation, and pulse rate sensors. Statistical analysis is working to check the system's efficiency [21].

It monitors the electrocardiogram (ECG) signal of older adults remotely for heart problem symptoms and sends an urgent notification to the doctor [22].

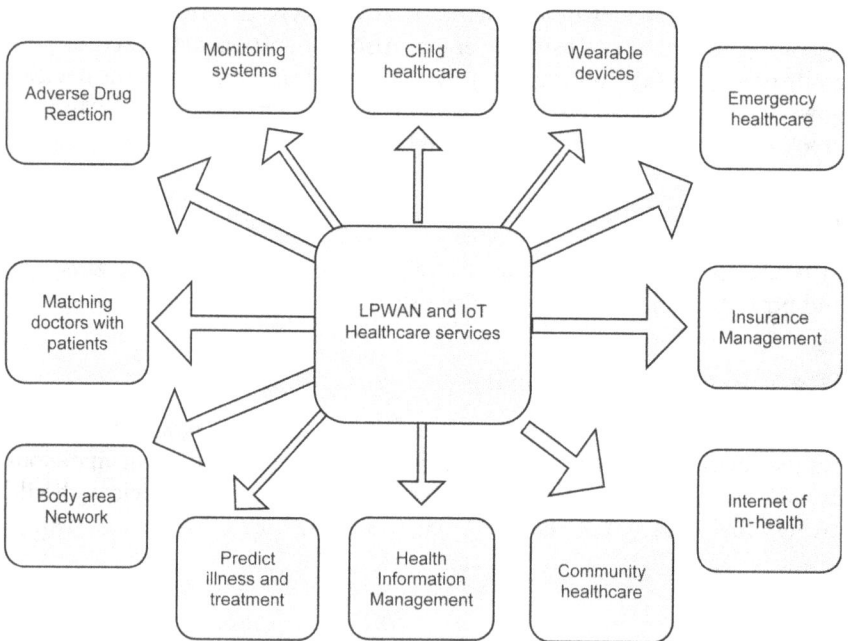

Figure 11.6 Healthcare services for LPWAN.

11.5 FUTURE OF HEALTHCARE IoT NETWORKS USING LPWAN

The number of IoT devices is increasing daily; there were 12.4 billion devices in 2020 [23], and the forecasted number is expected to have 26 billion IoT devices by the end of 2026. Figure 11.7 shows these figures based on the Ericsson Mobility Report (November 2022) [24].

LPWANs took over as the primary force behind the expansion of IoT connections globally in 2021 and are anticipated to eventually replace the majority of 2G/3G IoT connections [25].

The LPWAN environment is in constant flux because it is a relatively young technology and is still in its infancy. Telecommunication development is continuous, and every day, we have new technology or more advances in the existing ones, which are reflected in the health sector, facilitate medical management, and solve many issues.

Today we have body medical sensors and mobile computing, which create the concept of m-Health. Machine learning and artificial intelligence are applied in some situations to serve the health sector, especially during the pandemic, so it's a good experience to apply this experience in different areas of health sector needs [26].

Machine learning could also be used to predict the patient's behavior and the pattern of their healthcare needs [27].

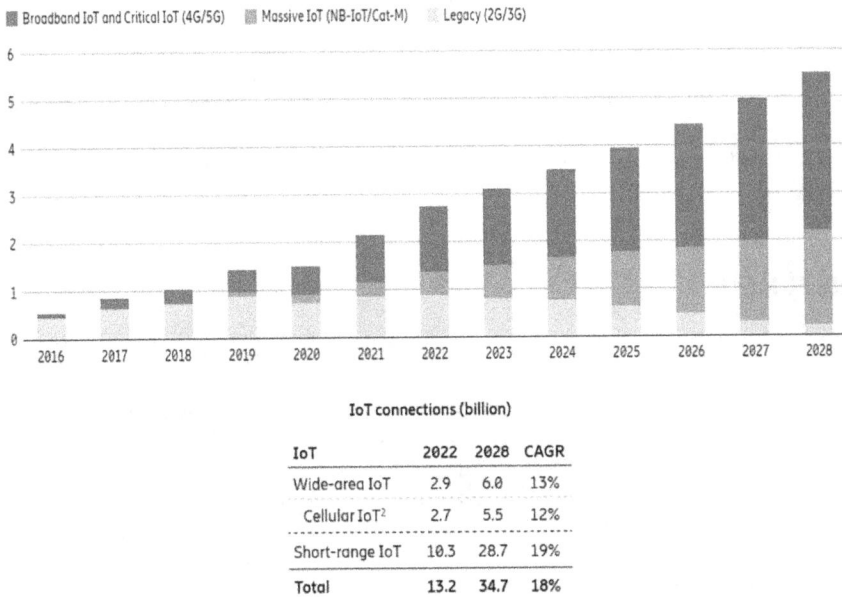

IoT	2022	2028	CAGR
Wide-area IoT	2.9	6.0	13%
Cellular IoT[2]	2.7	5.5	12%
Short-range IoT	10.3	28.7	19%
Total	13.2	34.7	18%

Figure 11.7 IoT devices forecast (cellular IoT connections by segment and technology billions).

Indeed, we are looking for more cooperation between these new technologies to make the patients' lives easier for them and the healthcare providers.

11.6 CONCLUSION

LPWAN technology has emerged as an exceptionally efficient and dependable solution for healthcare IoT networks. With its unique combination of low power consumption, long-range connectivity, and secure data transmission capabilities [28], LPWAN has positioned itself as a key enabler in revolutionizing healthcare practices. As the demand for IoT technology in healthcare continues to soar, LPWAN networks are expected to play a vital role in enhancing patient outcomes and transforming healthcare delivery.

By leveraging LPWAN technology, healthcare practitioners can connect various medical devices and sensors, facilitating real-time monitoring of patients' vital signs, medication adherence, and overall health status. The low power requirements of LPWAN devices ensure extended battery life, allowing for continuous monitoring without the need for frequent battery replacements. Moreover, the long-range connectivity offered by LPWAN enables the creation of expansive networks that cover large hospital campuses or even entire communities, ensuring seamless connectivity and data transfer across multiple locations.

One of the most significant advantages of LPWAN technology in healthcare IoT networks is its emphasis on data security [29]. Maintaining robust security measures is paramount with the sensitive nature of patient health information. LPWAN protocols employ encryption techniques and authentication mechanisms, safeguarding data transmission and protecting against unauthorized access. This ensures that patient data remains private and confidential, complying with stringent regulatory requirements such as HIPAA.

REFERENCES

[1] Albulayhi, K.; Al-Haija, Q. A.; Sheldon, F. T.: "Effective learning-based attack detection methods for the Internet of Things." (Computing, 2022). In *AIoT Technologies and Applications for Smart Environments*, Chap. 15. IET Digital Library, pp. 275–293, DOI: 10.1049/PBPC057E_ch15.

[2] Al-Haija, Q. A.; Jebril, N. A.: 'Time series data air quality prediction using Internet of Things and machine learning techniques." (Computing, 2022). In *AIoT Technologies and Applications for Smart Environments*, Chap. 8. IET Digital Library, pp. 127–150, DOI: 10.1049/PBPC057E_ch8.

[3] Janani, M.; Nataraj, M.; Shyam Ganesh, C. R.: "Mining and monitoring human activity patterns in smart environment-based healthcare systems." (2020). In *Systems Simulation and Modeling for Cloud Computing and Big Data Applications*. Academic Press, pp. 137–145.

[4] Al-Haija, Q. A.; Alnabhan, M.; Saleh, E.; Al-Omari, M.: Applications of blockchain technology for improving security in the Internet of Things (IoT). (2023). In *Blockchain Technology Solutions for the Security of IoT-Based Healthcare Systems*. Academic Press, pp. 199–221.

[5] Li, Y.; Chen, W.; Ding, Y.; Qie, Y.; Zhang, C.: "A vision of intelligent IoT - Trends, characteristics, and functional architecture." (2022). In *2022 International Wireless Communications and Mobile Computing (IWCMC)*, Dubrovnik, Croatia, pp. 184–189. DOI: 10.1109/IWCMC55113.2022.9824304.

[6] SpiceWork. "Top 10 Applications of IoT in 2022: The Internet of Things is a network of devices that feed data into a platform to enable communication and automated control." (May 2022). *Technical Blogs*. Retrieved online via https://www.spiceworks.com/tech/iot/articles/top-applications-internet-of-things/#_001 (Accessed on Feb.03.2023).

[7] Queralta, J. P.; Gia, T. N.; Tenhunen, H.; Westerlund, T.: "Edge-AI in LoRa-based health monitoring: Fall detection system with fog computing and LSTM recurrent neural networks." (2019). In 2019 42nd International Conference on Telecommunications and Signal Processing (TSP), Budapest, Hungary, pp. 601–604. DOI: 10.1109/TSP.2019.8768883.

[8] Aileni, R. M.; Suciu, G.; Sukuyama, C. A. V.; Pasca, S.; Maheswar, R. "Internet of wearable low-power wide-area network devices for health self-monitoring." (2020). In Bharat S. Chaudhari, Marco Zennaro (eds.), *LPWAN Technologies for IoT and M2M Applications*. Academic Press, pp. 307–325.

[9] ByteLab. "LPWAN technologies used in IoT." (Aug 2022). *Technical Blogs*. Retrieved online via https://www.byte-lab.com/lpwan-technologies-used-in-iot/ (Accessed on Feb.04.2023).

[10] Chilamkurthy, N. S.; Pandey, O. J.; Ghosh, A.; Cenkeramaddi, L. R.; Dai, H.-N.: "Low-power wide-area networks: A broad overview of its different aspects." (2022). *IEEE Access*, 10, 81926–81959. DOI: 10.1109/ACCESS.2022.3196182.

[11] Kabalcı, Y.; Ali, M.: "Emerging LPWAN technologies for smart environments: An outlook." (2019, June). In *2019 1st Global Power, Energy and Communication Conference (GPECOM)*, IEEE, pp. 24–29.

[12] Tsavalos, N.; Abu Hashem, A.: "Low power wide area network (LPWAN) Technologies for Industrial IoT applications." (2018). https://lup.lub.lu.se/luur/download?func=downloadFile&recordOId=8950859&fileOId=8950964

[13] Khan, Z. A.; Abbasi, U.; Kim, S. W.: "Machine learning and LPWAN based internet of things applications in healthcare sector during COVID-19 pandemic." (2021), *Electronics*, 10, 1615. DOI: 10.3390/electronics10141615.

[14] Olatinwo, D. D.; Abu-Mahfouz, A.; Hancke, G.: "A survey on LPWAN technologies in WBAN for remote healthcare monitoring." (2019 Nov 29). *Sensors (Basel)*, 19(23), 5268. DOI: 10.3390/s19235268.

[15] Manatarinat, W.; Poomrittigul, S.; Tantatsanawong, P.: "Narrowband-internet of things (NB-IoT) system for elderly healthcare services." (2019). In *2019 5th International Conference on Engineering, Applied Sciences and Technology (ICEAST)*, Luang Prabang, Laos, pp. 1–4, DOI: 10.1109/ICEAST.2019.8802604.

[16] Taleb, H.; Nasser, A.; Andrieux, G.; Charara, N.; Cruz, E. M.: "Energy consumption improvement of a healthcare monitoring system: Application to LoRaWAN." (1 April 1, 2022). *IEEE Sensors Journal*, 22(7), 7288–7299, DOI: 10.1109/JSEN.2022.3150716.

[17] Shahidul Islam, M.; et al.: "Monitoring of the human body signal through the internet of things (IoT) based LoRa wireless network system." (2019). *Applied Sciences*, 9(9), 1884.

[18] Droos, A.; Al-Haija, Q. A.; Alnabhan, M.: "Lightweight detection system for low-rate DDoS attack on software-defined-IoT." (2022). In *6th Smart Cities Symposium (SCS 2022), Hybrid Conference*, Bahrain, pp. 157–162. DOI: 10.1049/icp.2023.0388.

[19] Al-Sarawi, S.; Anbar, M.; Alieyan, K.; Alzubaidi, M.: "Internet of things (IoT) communication protocols: Review." (2017). In *8th International Conference on Information Technology (ICIT)*, Amman, Jordan, pp. 685–690. DOI: 10.1109/ICITECH.2017.8079928.

[20] Al-Haija, Q. A.: "Design and on-field testing of wireless sensor network-based air quality monitoring system." (2019). *JITCE (Journal of Information Technology and Computer Engineering)*, 3(2), 54–59.

[21] Chen, Y.; Sambo, Y. A.; Onireti, O.; Imran, M. A.: "A survey on LPWAN-5G integration: Main challenges and potential solutions." (2022). *IEEE Access*, 10, 32132–32149. DOI: 10.1109/ACCESS.2022.3160193.

[22] Sinha, R. S.; Yiqiao W.; Seung-Hoon H.: "A survey on LPWA technology: LoRa and NB-IoT." (2017). *ICT Express*, 3(1), 14–21.

[23] Abu Al-Haija, Q.; Zein-Sabatto, S.: "An efficient deep-learning-based detection and classification system for cyber-attacks in IoT communication networks." (2020). *Electronics*, 9, 2152. DOI: 10.3390/electronics9122152.

[24] Marais, J. M.; Abu-Mahfouz, A. M.; Hancke, G. P.: "A survey on the viability of confirmed traffic in a LoRaWAN." (2020). *IEEE Access*, 8, 9296–9311. DOI: 10.1109/ACCESS.2020.2964909.

[25] Ali, Z.; Qureshi, K. N.; Al-Shamayleh, A. S.; Akhunzada, A.; Raza, A.; Butt, M. F. U.: "Delay optimization in LoRaWAN by employing adaptive scheduling algorithm with unsupervised learning." (2023). *IEEE Access*, 11, 2545–2556. DOI: 10.1109/ACCESS.2023.3234188.

[26] Krishnamurthy, I.; Prabha, D.; Karthika, M. S.: "IoT-based smart mirror for health monitoring." (2020). In J. Dinesh Peter, Steven L. Fernandes (eds.), *Systems Simulation and Modeling for Cloud Computing and Big Data Applications*. Academic Press, pp. 115–121.

[27] The 3G4G "Broadband IoT (4G/5G) connections to dominate by the end of 2028", (Nov 2022). Retrieved online via (Accessed on Mar.08.2023) https://www.rcrwireless.com/20230622/internet-of-things-4/total-iot-connections-to-surge-but-cellular-iot-to-lose-share-with-2g-3g-switch-off

[28] Moon, B.-K.: "Dynamic spectrum access for internet of things service in cognitive radio-enabled LPWANs." (2017). *Sensors*, 17, 2818. DOI: 10.3390/s17122818.

[29] Odeh, A.; Keshta, I.; Al-Haija, Q.A.: "Analysis of blockchain in the healthcare sector: Application and issues." (2022). *Symmetry*, 14, 1760. DOI: 10.3390/sym14091760.

Chapter 12

Internet of things-based smart healthcare application using LoRa network

G. Victo Sudha George, L. Dharani,
Billa Hemalatha, and Bommidala Abhishake
Dr. M.G.R Educational and Research Institute

12.1 INTRODUCTION

The growth of the healthcare sector as an industry is being accelerated by technological breakthroughs. In the healthcare industry, the Low-Power Wide-Area Network (LPWAN) is a critical IoT enabler. Devices may connect to one another over extremely long distances using the wireless LPWAN protocol. The healthcare sector is one of the most important ones where Internet of Things (IoT) technology is having a big influence on patient outcomes and the delivery of better healthcare services. The use of IoT in healthcare enables remote patient monitoring and administration in addition to real-time data collecting from a range of medical devices, wearables, and sensors. A communication technology called Wide-Area Low-Power Network is revolutionizing IoT in the healthcare industry. LPWAN enables low-power data transport over great distances, enabling remote patient administration and monitoring. In this article, we'll look at how LPWAN and IoT are changing the healthcare industry.

12.2 FUNDAMENTALS OF IoT

In this section, the basic concepts behind IoT are discussed.

12.2.1 What is an IoT?

Low-cost computer processors and high-bandwidth connections led to the development of the IoT, a network of linked things connected to the internet. This enables technology, such as automobiles, robots, and vacuum cleaners, to collect data and respond to customers. The IoT connects common place "things" to the internet. Computer developers gradually incorporated sensors and processors into everyday objects during the 1990s. RFID tags were first employed to track expensive equipment, but as computer systems got smaller, processors got faster, smarter and smaller. A network of physical

DOI: 10.1201/9781003426974-12

objects that have been equipped with sensors, software, and other technologies to enable data collection and exchange with other systems and devices is known as the "Internet of Things" (IoT). The three main types of IoT hardware are consumer, industrial, and commercial. IoT may be used for many things, including enhancing human well-being and increasing company productivity and efficiency. Examples of popular use cases include smart home automation, healthcare monitoring, supply chain management, environmental monitoring, and others. The IoT is a rapidly developing topic because it offers so many chances to improve our lives and fundamentally alter how we interact with technology. The working principles of IoT are given in Figure 12.1.

IoT devices may be divided into three groups:

Devices like wearable technology, smart home appliances, and personal health monitoring devices are all part of the consumer IoT.

Industrial IoT includes gadgets used in transportation, agriculture, and manufacturing. These tools can help in streamlining procedures, lowering expenses, and increasing output [1].

Technology used in a range of industries, such as hospitality, retail, and healthcare, is referred to as "commercial IoT." Intelligent building automation systems and location-based services are two examples (Figure 12.2).

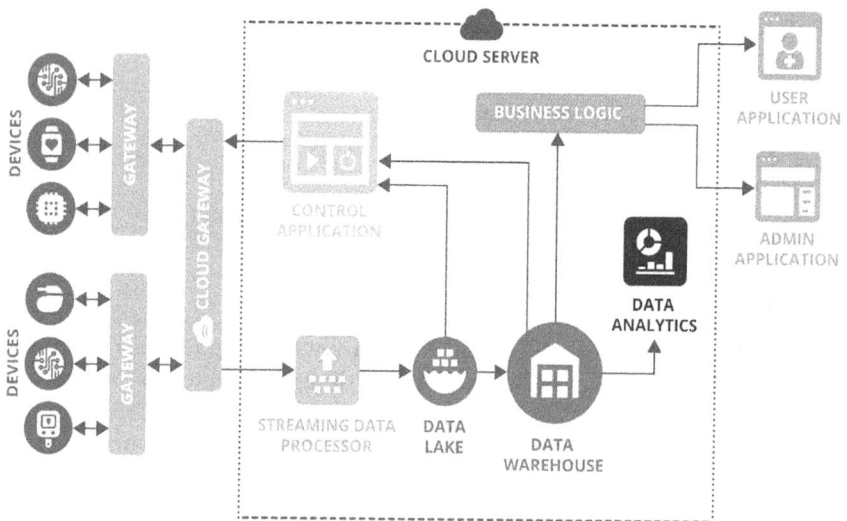

Figure 12.1 Working of IoT.

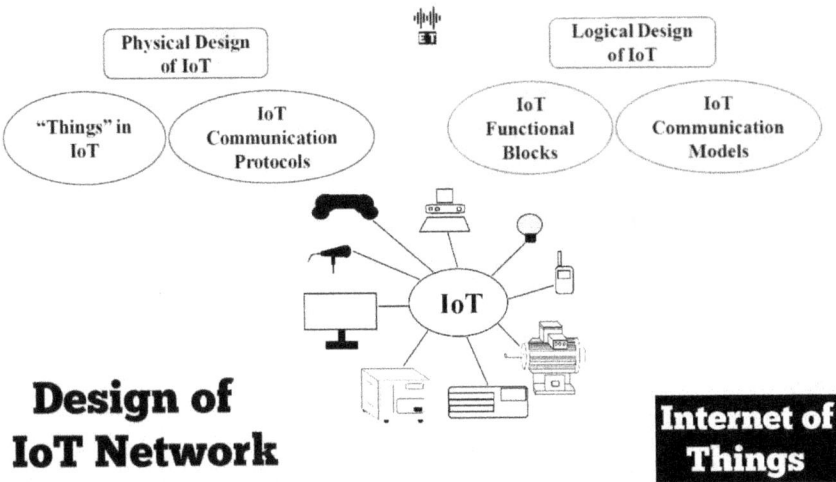

Figure 12.2 Design of IoT networks.

12.2.2 IoT communication standards

The two categories for IoT communication protocols are LPWAN and short-range networks (ICIT), shown in Figure 12.3.

Wide Area Low Power Network (LPWAN): SigFox is a low-power alternative for M2M applications and wireless sensor transmission. To transmit

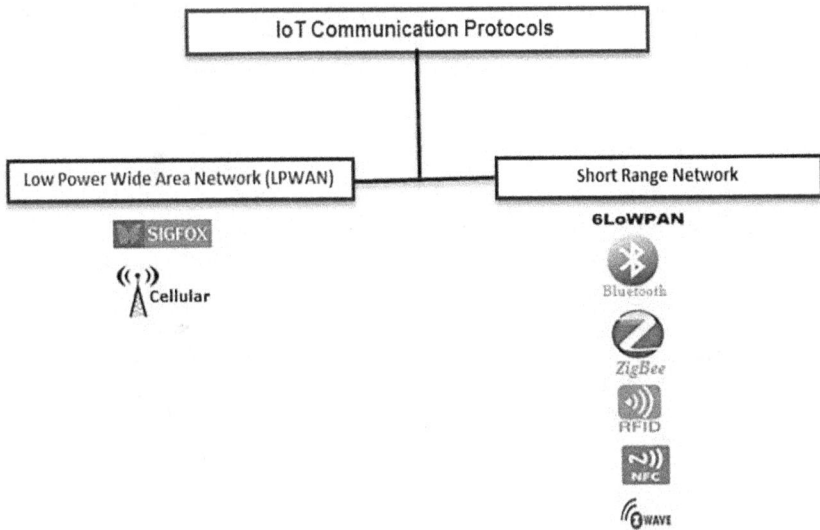

Figure 12.3 IoT communication protocol.

tiny quantities of data up to 50 km, SigFox employs UNB technology. Smart metres, patient monitoring, agricultural equipment, security gadgets, street lights, and environmental sensors all make use of NFC technology [2]. It is powered by a small battery and built to endure slow data transfers. The initial network's topology is supported by SigFox.

Cellular applications that require high data rates and IoT applications that require a power source are the greatest candidates for cellular technology. It has GSM, 3G, and 4G cellular connectivity capabilities; however, it uses a lot of electricity. M2M or local network communication is unacceptable, even though it is employed for numerous applications. Many forms of technology are based on cellular structure.

12.2.3 Uses of IoT

IoT may be utilized for a multitude of purposes, including enhancing people's quality of life and increasing company productivity and efficiency. Common IoT use cases are:

- **Smart home automation:** The lighting, temperature, and security of a home may all be controlled and automated with the aid of IoT devices.
- **Healthcare monitoring:** IoT devices may be used to keep tabs on patients' health states and share information with medical staff, allowing for remote monitoring and faster diagnosis [3].
- **Supply chain management monitoring:** To maintain stocks, monitor equipment, and optimize shipping routes while increasing efficiency and reducing costs, supply chain management may deploy IoT devices.
- **Environmental monitoring:** IoT gadgets may be used to keep track of weather patterns, search for toxins, and evaluate the water and air quality.

12.2.4 What is NB-IoT

The NB-IoT is an LPWAN technology that permits connections between devices. To make 3G and 4G/LTE communication possible, the 3GPP developed a network of mobile phone signals [4]. NB-IoT devices with modest data rates and battery usage use cellular networks.

Standalone, guard-band, and in-band are its three operational modes. The most significant element of NB-IoT is that it can be used in two ways: guard-band, which functions like a typical wireless sensor, and in-band, which makes use of the usual LTE frequency. Guard-band maintains frequency channels apart from one another to prevent interference, in contrast to in-band, which takes use of open space.

- Depending on the cell site and base station that the device is supported by, NB-IoT offers benefits including enhanced connection, flexibility, and cost savings

- Battery life of 10 years, 100,000 IoT devices per base station, and low power requirements for uplink data
- The network's coverage area may go up to 10 km, allowing for crystal-clear signal reception inside of structures
- Constructing a network using a co-location system to speed up the rollout of IoT services
- A gyroscope sensor is used to detect the rotational characteristics of a smartphone using three-axis sensing. To respond accurately and quickly to user motions, smartphones also come equipped with an accelerator sensor. No matter how the smartphone is used, the combination feature provides precision and control

12.2.5 How do the IoT work?

IoT connects physical objects, devices, and sensors to the internet in order to gather and exchange data with other systems, devices, and sensors [5]. The architecture of IoT is given in Figure 12.4. This is a brief description of the operation of the IoT:

- **Devices and sensors:** IoT devices, such as smart thermostats, fitness trackers, and security cameras, have sensors built in that collect data about the environment, such as position, motion, and temperature
- **Connectivity:** IoT devices employ a range of connection methods, such as Wi-Fi, Bluetooth, and cellular networks, to communicate data over the internet
- **Data processing:** IoT device data is processed and analysed using software and algorithms to get insights and identify patterns
- The conclusions drawn from the data inform IoT activities and reactions. Users can acquire and exchange data in this way

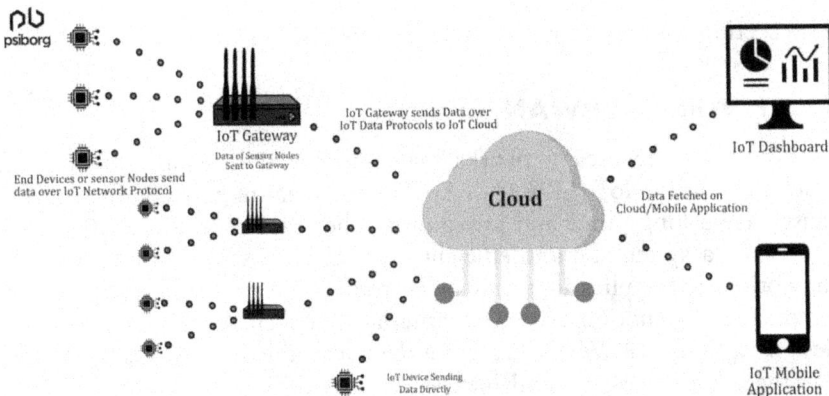

Figure 12.4 Architecture of IoT.

- **Cloud computing:** To store, process, and analyse data from many sources, IoT devices make use of cloud computing platforms
- Tethering physical objects, external devices, and sensors is how the IoT functions
- **Actions and Reactions:** Based on the information gleaned from the data, IoT devices may automate processes, such as changing a room's temperature, or sending people notifications or messages [6].

Many academicians have published numerous designs. But no IoT design has gained general approval. Although three-layer structures with layers for sensing, networking and applications are the most popular in this industry, five-layer structures are also employed.

For information gathering and sensing, the physical layer is equipped with sensors.

i. The perception layer is made up of ambient data. It differentiates itself from other environment-based smart objects, recognizes physical qualities, detects a change in physical attributes, or both.
ii. The network layer is in charge of establishing connections with servers, networking equipment, and other intelligent things. It uses its traits to send and process sensor data.
iii. The application layer is in charge of providing the user with services tailored to certain applications, such as smart homes, smart cities, and smart health.

The IoT's three-layer architecture is inadequate and calls for the deployment of more layered systems. The layered architecture of IoT is given in Figure 12.5.

12.3 LPWAN

This section speaks about the fundamentals and the features of LPWAN.

12.3.1 What is LPWAN, exactly?

Low power wide area networks may improve the connection between machines and the IoT (LPWAN). Applications that require communication must have a long range and low power. The main objective of this work is to offer a systematic and efficient method for discovering the underlying notion that applications may be customized to particular applications, expressing that notion as precise criteria, and then selecting the relevant design options. LPWANs are resource-constrained networks with vast coverage, high capacity, and inexpensive device and installation costs as their primary features [7]. These characteristics result in a sizable set of

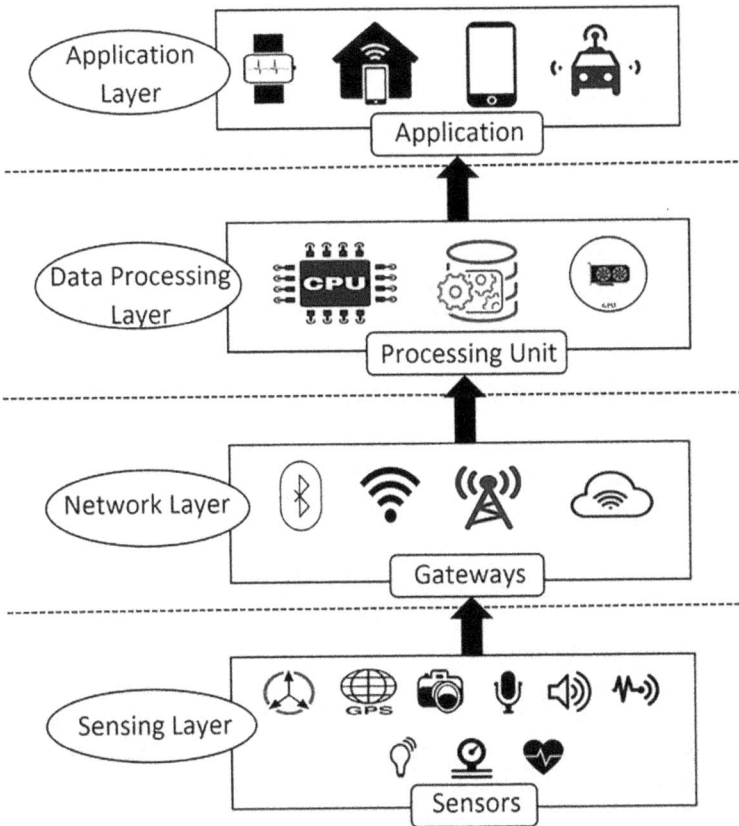

Figure 12.5 Different layers in IoT.

requirements, which include M2M traffic management, massive capacity, energy efficiency, low power operations, better coverage, security, and interoperability.

The permitted design considerations are grouped into two groups: desired or expected and enhanced, to represent the wide variety of characteristics associated with LPWAN-based applications. Precise location identification, security coverage techniques, lightweight media access control (MAC) protocols, interference management, energy-saving modes of operation, and adaptive software reconfigurability are some of the most crucial design elements. The topological and architectural options for LPWAN entity connection are discussed. The most popular proprietary and standard-based commercial LPWAN technology solutions are discussed. Sigfox, LoRaWAN, NB-IoT, and long-term evolution LTE are a few of them. The promise of cellular 5G technology and how it enhances LPWAN technology are also discussed. The description of LPWAN is given in Figure 12.6.

Figure 12.6 Description of LPWAN.

With the help of Low-Power Wide-Area Network (TLPWAN) technology, devices may communicate wirelessly over extremely great distances while using very little power.

This is a brief description of how LPWAN technology works:

- **LPWAN devices and sensors:** These devices use sensors, metres, or tracking devices to obtain data about the physical world, such as temperature, humidity, or location [8].
- **Connectivity:** LPWAN devices use a low-power wireless communication protocol like LoRaWAN, SigFox, NB-IoT, or LTE-M to send the collected data over long distances, sometimes several kilometres or more, to a gateway or base station [9]. Data from the LPWAN devices is received by the gateway or base station, which then sends it through a wired or wireless network to a centralised server or cloud platform.
- **Central server or cloud platform:** Software and algorithms are used to process and analyse the data collected by the LPWAN devices in order to get insights and identify patterns.
- **Actions and reactions:** Based on inferences drawn from the data, LPWAN systems may automate processes like altering the temperature of a room or sending users warnings or notifications.

12.3.2 The design and major features of LPWAN networks

LPWANs collect data using a star topology from spread out, far-off end nodes. The base station receives the messages and demodulates them before sending them across a TCP/IP backhaul link to the backend server. Private LPWANs enable data to be transported directly to the user's chosen back-end for complete data privacy and control, in contrast to publicly managed systems, which mandate that data be routed through the network opera-tors' server before reaching the end user's apps [10]. Large packet head-ers are therefore avoided (e.g., NB-IoT).What makes LPWAN so appealing are its long range and low power consumption, two advantages that were routinely sacrificed in earlier systems. LPWANs may broadcast signals far further than Wi-Fi and Bluetooth, up to 15 km in rural regions and 5 km in heavily crowded metropolitan areas. Moreover, this wireless family's deep penetration capabilities allow for the connecting of devices in inaccessible subterranean and interior locations. Moreover, it includes simple, small transceiver architecture to reduce costs and power consumption on the end node side. The objective is to shrink the data frame size while delegating all labor-intensive tasks to the base station. The comparison of LPWAN's power efficiency against other networks is given in Figure 12.7.

The IoT, which is now undergoing rapid growth, is projected to bring next-generation services like asset tracking and machine-to-machine (M2M) communication. The economic activities of human civilization include the healthcare sector. The IoT in healthcare is anticipated to deliver the highest level of service while cutting treatment costs, minimizing device downtime and supply replenishment, and enhancing patient satisfaction. Applications include ECG monitoring, hospital supply tracking, and ambient assisted living. The two categories of IoT healthcare apps are clustered applications, which cure a variety of ailments or diseases, and standalone applications, which concentrate on a particular service. IoT applications may be catego-rized into three types according to the amount of the coverage area they can

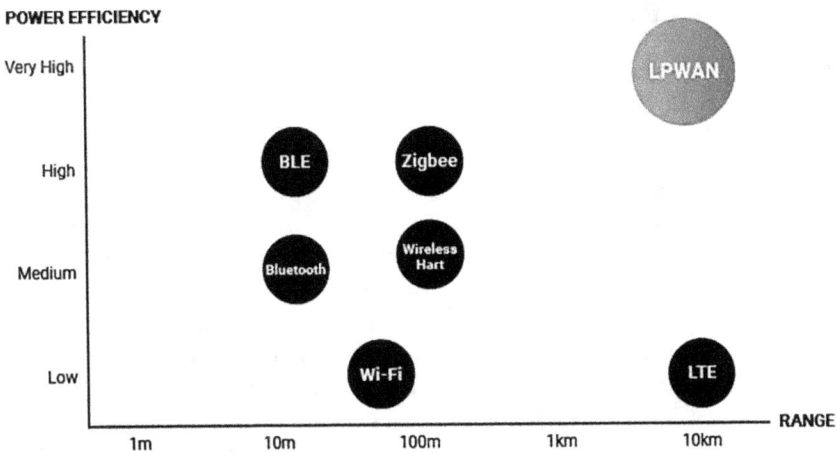

Figure 12.7 Compares the LPWAN's power efficiency against others.

provide. Bluetooth, ZigBee, and Z-Wave are a few examples of short-range wireless networks that enable communication up to a few metres. WLANs are Wi-Fi networks that allow for communication across a range of a few tens of metres or less. Mobile communications can evolve more easily over time thanks to cellular networks. The market is increasingly enticing to manufacturers since these well-established networks can connect with IoT devices across networks. Wide coverage, high network capacity, low data rates, low power consumption, and economical equipment are the primary criteria for IoT health applications. Initiatives to increase the range of WLANs and LPWANs or decrease the cost and power requirements of cellular networks led to the development of LPWA. LoRa is a wireless communication technology that was created to support M2M and IoT applications that use less power. Similar to SigFox and WiSun, LoRa is a promising and popular LPWA technology. LoRa employs spread spectrum modulation, a communication technique that was created using CSS technology. Using the 868 GHz ISM band, LoRa can provide coverage for up to 15 km on land and 30 km on water. The highest transmission power of LoRa, which has several mobile network providers, is 14 dBm. Basic communication capabilities are provided by the LoRaWAN communication architecture. The LoRaWAN protocol is used by endpoints that require batteries when combined with LoRa technology. The network is organized using a star-of-stars layout, with gateways connecting endpoints to the main network server. End devices and gateways can communicate with one another via LoRa or FSK. Network servers and LoRa Gateways are connected by standard Internet Protocol (IP) connections. End devices choose channels in a pseudo-random manner while using LoRa, which supports data speeds of up to 50 kbps, to reduce interference.

12.3.3 IoT LPWAN usage

LPWAN technology may be used by IoT applications in a variety of ways, depending on the specific use case and application. Some common IoT use cases for LPWAN include the following

- **Asset tracking:** With LPWAN technology, it is possible to follow the location of various assets across a wide region, including vehicles, containers, and machinery. Applications in supply chain management and logistics can truly profit from this
- **Smart cities:** LPWAN may be used to connect sensors and equipment that are a part of the infrastructure of a smart city, such as traffic lights, parking metres, and waste management systems. This could increase efficiency while lowering costs for public services
- **Agriculture:** In agricultural applications, LPWAN may be used to monitor soil moisture, temperature, and other environmental conditions.

Other environmental factors might also be present. By boosting agricultural output while using less water, farmers may gain from this

- **Industrial monitoring:** LPWAN may be used to monitor machinery and equipment in industrial settings like factories and warehouses. This can boost output, reduce downtime, and prevent equipment breakdowns
- **Healthcare:** The use of LPWAN in healthcare applications enables both the remote monitoring of patients' vital signs and the transfer of data to healthcare specialists. As patient care is enhanced, costs can be reduced

12.3.4 LoRa system

The two main components of the proprietary LPWAN technology developed by the LoRa Alliance are LoRa and a LoRaWAN protocol. In wireless networks like WBANs, LoRa, which stands for long range and is a proprietary wireless technology, may be used to target low-powered gadgets and devices that don't require a lot of data at once. Hence, LoRa might be used to expand network capacity, achieve a wide communication range, lower the cost of devices, and prolong battery life. The LoRa PHY layer employs either ISM bands 868 or 915 MHz depending on the deployment location. A single LoRa BS or gateway may be able to offer coverage over a 100-km metropolitan region, depending on the surroundings and other variables. Compared to other outdated communication technology standards, LoRa has a larger link budget. Excellent examples of sensing and monitoring systems that incorporate LoRa technology are smart metres, WBAN systems in healthcare, and WSN systems for environmental monitoring. Wireless communication systems are made possible by LoRa's low-power and long-range communication characteristics. The LPWAN layers are depicted in Figure 12.8.

The LoRaWAN wide-area network protocol, which is implemented at the MAC layer, builds a network architecture using the PHY layer of LoRa technology. To increase battery life during long-distance communication, the protocol makes use of ALOHA modulation and a star network architecture. The publications are examples of research investigations that considered employing LoRa for HCM. According to the aforementioned research, a LoRa solution may be included in remote HCM, but there are a number of drawbacks, such as a small network size, a poor data transmission rate, a high installation cost, and interference issues. The LoRa Gateways and backhaul systems use IP to link the sensor devices to the network server. The LoRa architecture consists of a remote device with a computer, a LoRa gateway, a server, and software. Medical staff may use the hospital database server to get patient health data, deliver care, and carry out other duties.

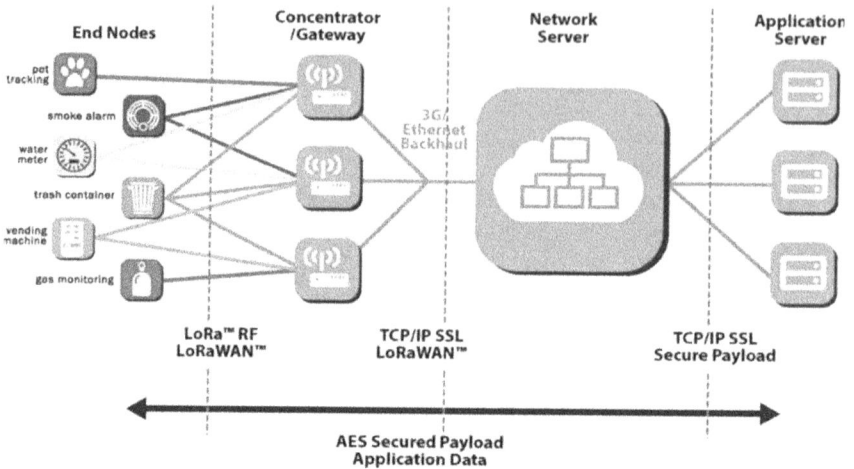

Figure 12.8 Displays several LPWAN layers.

12.4 USE OF LPWAN IN IoT HEALTHCARE

Machine learning is used in IoT-based healthcare applications to enhance healthcare services. These programs include ones for tracking blood pressure, ECGs, and glucose levels. As blood glucose monitoring reveals a person's blood glucose trend, it is essential for those with diabetes. ECG monitoring is used to evaluate various arrhythmias, cardiac rhythm, and myocardial ischaemia. Using LPWAN for blood pressure monitoring is crucial for many people. Monitoring blood pressure is essential for spotting many heart problems early on.

The architecture suggested for monitoring blood pressure utilizing IoT and smartphones, LPWAN devices, machine learning, and pulse oximetry is the most crucial piece of information in this study. Even if machine learning may be used to build a model for the prognosis of a prospective illness, the design gives a communication channel for continuous blood pressure monitoring between a remote health office and the main medical institution. The data gathered from these applications may also be used by a machine learning software program to spot any possible health problems in the subject being watched. Rehabilitation management is crucial in helping persons with physical impairments regain their functional capacity.

12.4.1 The MIA and mHealth network

A network of internet-connected hardware and software program is used to collect, organize, and share health data.

For the Body Area Network (BAN) component of the mHealth domain, a variety of end-to-end connection segments are developed, including data collecting, information management, user-centric device/sensor capabilities, protocol, and data exchange [11]. Other BAN variants include WBAN, Body Sensor Networks (BSN), and Medical Body Area Networks (MBAN). The IoT, a characteristic that makes many devices internet-aware, needs the development of certain skills in order for users and patients to properly route health-related data and get feedback from cloud and internet stakeholders. Patients and consumers now have simpler access to healthcare services as a consequence.

IoT is applied in healthcare to tackle specialized and widespread health problems. A few instances of isolated circumstances are electrocardiography, blood pressure, oxygen saturation, and body temperature. Several of them asserted that the two main applications of IoT in healthcare were single health problems and cluster health (more than one condition) difficulties, including wheelchair, treatment, and smartphone intervention apps for cluster disorders.

Healthcare alternatives IoT-based healthcare services are offered by the Internet of mHealth Things (mIoT), mHealth Community Healthcare (mCH), and mHealth Paediatric Healthcare. The interconnection of IoT devices with systems and designs found in urban infrastructure is commonly referred to as a "smart city." The goal of the smart city paradigm is to properly manage devices and associated data in order to monitor and analyse stakeholder interactions in metropolitan areas. The Figure 12.9 explains the application of LPWAN in various domains.

Based on device and health data that has been analysed by an inference algorithm, smart devices are equipped with the capability to raise an alert and send it to a regional headquarters. To reduce false alarms, smart gadgets

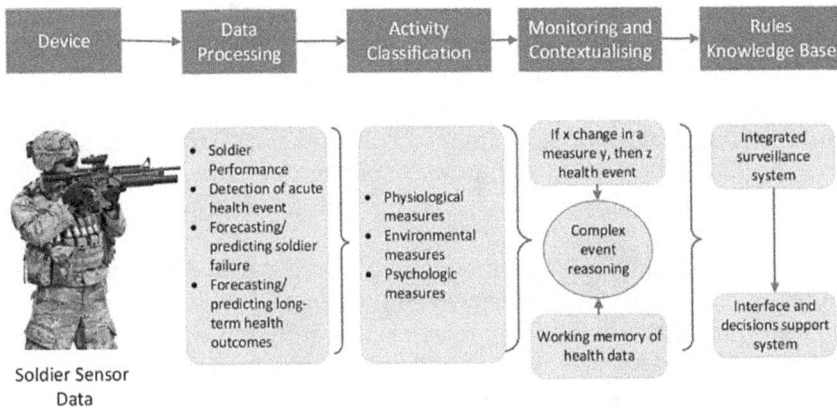

Figure 12.9 Application of LPWAN.

request permission before sounding an alert. A timeout is enforced and an alarm is automatically set off if a solution cannot be discovered in the specified time. To determine if the user's condition necessitates an alarm, the predefined threshold table employs a scenario determination technique. In order to identify if a person is conscious, wounded, or both, activity recognition is utilised to recognize their posture or motions.

A Libelium eHealth shield, an Arduino Mega microcontroller development board, and a USB power bank were included with the LoRaWAN health monitoring system. The system successfully sent data on temperature, glucose, systolic and diastolic blood pressure, and pulse rate with gateway coverage ranging from 0.73 to 1.89 km. Due to the system's sluggish transmission rate, ECG transmission was not enabled. For HRV analysis, a LoRa node running the LOMB algorithm is connected to the cloud using LoRaWAN. Inside the 3-km gateway coverage range of the LoRa node are a PPG for pulse rate detection and a DSP for HRV analysis.

A data inference method for mHealth security is provided by the mHealth network to reduce data transmission frequency and protect sensor device battery life. By including an optimisation stage, MIS may be more accurate and productive.

Biometrics and health data together improve multi-factor authentication.

12.4.2 Ethos and exercise scenarios

This section covers the use of the inference system to optimize the data set and decrease data size. As was already indicated, several studies have looked at a range of inference techniques, including the employment of beacons, the application of variance rate, and the elimination of duplicate data. A good inference system can properly represent the original data source with fewer data extractions. In this experiment, the MIS could assess the variance of sensed data and compare it to predefined threshold values. A health expert determines the threshold after assessing the user and taking into account their age, gender, health conditions, and status.

The four main physiological measures are blood pressure (BP), heart rate (HR), and body temperature (BT) (BP). Due to its fast adaptability and sensitivity to changes in physical and psychological state, HR is better suited to testing and the creation of enormous amounts of data. It was chosen to be used in the study's inference system because it can be made fast in large quantities.

The main objectives of the inference system are to reduce energy consumption and transmission frequency in sensor nodes, both of which are crucial in military applications. By refraining from uploading unmodified or essentially comparable data as well as data irrelevant to the data analysis stage, military systems can deploy such an inference system. In this work, the HR data set's highest and lowest HR data points—or the relative extrema—are optimized in two layers utilizing an MIA. By utilizing several layers and phases of data optimization, the MIA can decrease the amount of data samples from the original data set.

To collect data samples, the MIA divides predetermined durations into short and long intervals depending on whether the user is exercising, using regular mode, or not exercising at all. Figure 12.10 explains the interaction amoung the different layers.

To be effective and have a high rate of savings, the MIS developed for this project has two inference layers and a total of four phases.

1. The inference system's first layer retrieves information from the first data gathering.
    a graph showing how the true value and the implied value are related.
2. The accuracy increases when the savings rate decreases because the first layer's distance between the real and inferred data points is reduced.
3. Data from the earlier inferred data collection are used in the second layer of the inference system, which decreases accuracy while boosting savings.
4. The second layer adds more valuable data to the inferred data set, improving the extracted data set from Step 3 and improving algorithm accuracy. Precision is improved by this process. Hence, even if the

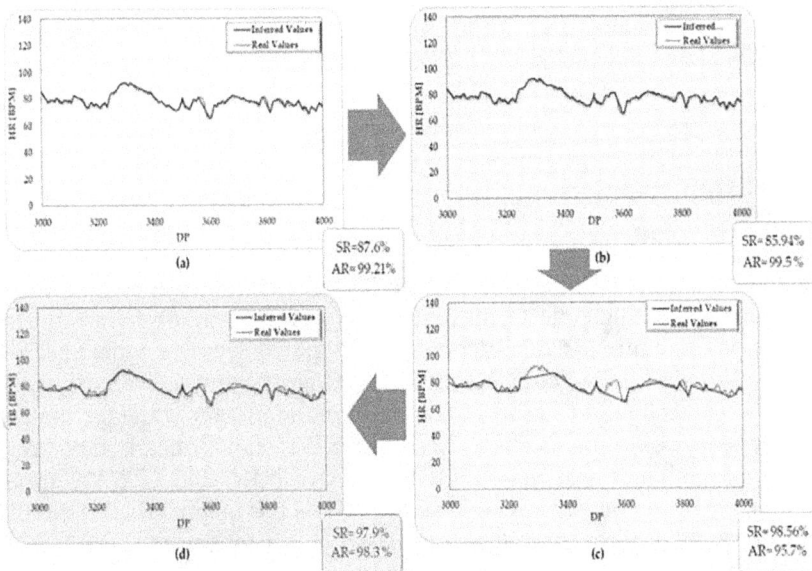

Figure 12.10 Interaction between layer.

savings rate is slightly lower, the accuracy rate significantly rises. The savings rate increased by 10.3% in comparison to the baseline results and the inferred results using 1,000 data points; however, the accuracy rating decreased by 0.91%.

LPWAN is used by IoT in healthcare in a number of ways, including:

By the use of LPWAN-enabled medical devices and wearables [12] patient data may be collected and communicated in real-time to healthcare experts, enabling remote patient monitoring of patients outside of traditional healthcare facilities.

Asset management and tracking: By keeping track of their tools and supplies, healthcare facilities may reduce waste and increase productivity. LPWAN support for IoT makes this possible.

By keeping an eye on the performance of medical equipment and notifying clinicians when repairs are necessary, predictive maintenance may save downtime and enhance patient care. Healthcare professionals may treat patients who are located far away by employing remote diagnostic equipment and video conferencing that are LPWAN-enabled, improving patient access to services.

Emergency response: LPWAN-enabled alarm systems can alert medical personnel to situations requiring immediate attention, such as heart attacks or accidents. This might lead to the administration of potentially life-saving medication.

All things considered, the healthcare industry may gain from IoT using LPWAN technology by improving patient care, increasing operational performance, and reducing costs [13].

12.5 CONCLUSION

The use of IoT and LPWAN technologies may have an impact on the healthcare industry. By connecting medical devices, wearables, and other equipment to a single platform, healthcare workers may optimize clinical operations, enhance patient outcomes, and remotely regulate patient health. The long-distance, low-cost, and low-power networking solution provided by LPWAN technology ensures data security and privacy while facilitating seamless device communication. Interoperability, standardization, and regulatory compliance challenges must be resolved if IoT and LPWAN are to completely benefit healthcare. IoT with LPWAN has the potential to totally transform the healthcare industry by offering patient-centred, preemptive treatment, improved operational performance, and cheaper healthcare costs. The use of IoT with LPWAN in healthcare might be advantageous based on the advantages that this technology may provide to the industry. IoT with LPWAN enables medical staff to remotely monitor and manage patient health, speed up clinical processes, and enhance patient outcomes by connecting medical devices, wearables, and other equipment to a single platform. LPWAN technology offers a low-cost, low-power, and long-range connection solution that makes it feasible to offer patients individualized and proactive therapy in addition to protecting data confidentiality and privacy. Notwithstanding challenges like regulatory compliance, standardization, and interoperability, utilizing IoT with LPWAN in healthcare has a lot of potential advantages.

REFERENCES

1. B. Sabin, M. Radu Jr, R. Alexandru, "Wearable IoT power consumption optimization algorithm", Technical University of Cluj-Napoca, 2020 International Conference and Exposition on Electrical And Power Engineering (EPE). doi: 10.1109/EPE50722.2020.9305609.
2. S. Al-Sarawi1, M. Anbar, K. Alieyan, M. Alzubaidi, "Internet of Things (IoT) Communication Protocols", National Advanced IPv6 Centre (NAV6), Universiti Sains Malaysia, 2017.
3. J.J. Kang, W. Yang, G. Dermody, M. Ghasemian, S. Adibi, P. Haskell-Dowland, "No soldiers left behind: An IoT-based low-power military mobile health system design", *IEEE Access* 8:201498–201515. doi: 10.1109,2020.3035812.
4. W. Manatarina, S. Poomrittigul, P. Tantatsanawong, "Narrowband-Internet of Things (NB-IoT) System for Elderly Healthcare Services", Pathumwan Institute of Technology Bangkok, 5th International Conference on Engineering, Applied Sciences and Technology (ICEAST). doi: 10.1109/ICEAST.2019.8802604.2019.
5. G. Panagi, K. Katzis, "Lead Electrocardiogram Monitoring Over Lora: A Conceptual Design", IEEE International Conference on Communications 2020 (IEEE ICC 2020). doi: 10.1109/ICCWorkshops49005.2020.9145424.
6. S. Al-Sarawi1, M. Anbar, K. Alieyan, M. Alzubaidi, "Internet of Things (IoT) Communication Protocols", National Advanced IPv6 Centre (NAV6), Universiti Sains Malaysia, 2017.
7. J. Pena Queralta, T. N. Gial, H. Tenhunen, T. Westerlund, "Edge-Al in Lora-Based Health Monitoring: Fall Detection System with Fog Computing and LSTM Recurrent Neural Networks", 2019 42nd International Conference on Telecommunications and Signal Processing. doi: 10.1109/TSP.2019.8768883.
8. P.V. Dudhe, N.V. Kadam, R.M. Hushangabade, "LPWAN for IoT", 2017 International Conference on Energy, Communication, Data Analytics and Soft Computing (ICECDS). doi: 10.1109/ICECDS.2017.8389935.
9. S.C. Gaddam, M.K. Rai, "A Comparative Study on Various LPWAN and Cellular Communication Technologies for IoT Based Smart Applications", 2018 International Conference on Emerging Trends and Innovations in Engineering and Technology. doi: 10.1109/ICETIETR.2018.8529060.

10. J. Petajajarvi, K. Mikhaylov, M. Hamalainen, Centre for Wireless Communication Engineering University of Oulu, Finland, "Evaluation of LoRa LPWAN Technology for Remote Health and Well Being Monitoring", *2017 International Journal of Wireless Information Networks*. doi: 10.1007/s10776-017-0341-8.

11. W. Manatarinat, S. Poomrittigul, "The Narrowband-Internet Of Things (NB-IoT) System For Elderly Healthcare Services", 2019 5th International Conference on Engineering, Applied Sciences and Technology (ICEAST). doi: 10.1109/ICEAST.2019.8802604.

12. U. Raza, P. Kulkarni, M. Sooriyabandara, "Low power wide area networks: An overview", *IEEE Communications Surveys & Tutorials*, 19(2): 855–873, 2017.

13. R. Sanchez-Iborra, M.-D. Cano, "State of the art in LP-Wan solutions for industrial IoT services", *Sensors*, 16(5): 708, 2016. doi: 10.3390/s16050708.

Chapter 13

Deep-learning enabled early detection of COVID-19 infection in IoMT fog-cloud healthcare in LPWAN

Mujeeb ur Rehman, Rehan Akbar,
Shagufta Mujeeb, and Aftab Alam Janisar
Universiti Teknologi PETRONAS, Malaysia

13.1 INTRODUCTION

In December 2019, the COVID-19 epidemic began in Wuhan, China, and has since spread throughout the world. The virus has been associated with severe respiratory illness, resulting in a high number of hospital admissions, particularly in acute care units, as well as a significant number of deaths. In response, medical companies have been working tirelessly to develop test kits that can accurately detect and diagnose COVID-19 [1]. However, the method used to take samples from the mouth and nose may have an impact on how accurate the test results are. This study focuses on using pictures of chest X-rays to create and train a deep learning model for the early identification of COVID-19 [2,3], despite the fact that CT scans of the lungs have shown some promise in aiding in the early diagnosis of coronavirus infection. The main method for classifying and differentiating between COVID-19 positive and negative patients is convolutional neural networks (CNNs) [4]. The COVID-19 picture data collection, an open-source dataset made accessible on the GitHub platform, was used by the researchers to train the deep learning model. Images from chest X-rays taken of COVID-19 positive and negative patients are included in the dataset. Before developing and training the deep learning model, the researchers preprocessed the dataset to improve its quality. They then created, trained, and tested the model, using subject-related specialized terminology, which they defined and explained in the methodology section. To develop an effective deep learning model for the early detection of COVID-19, the study limited the model to 130 patients' data [5].

Furthermore, Artificial Intelligence (AI)-based systems have the ability to predict epidemiological patterns due to their self-learning capabilities. Forecasting relies on past clinical data, but conventional medical systems only use hospital management systems to manage patient health records [6]. This creates a large amount of data that remains largely unused and requires effort to make intelligent decisions [7]. Using AI techniques can help diagnose diseases by reducing data ambiguity and providing better decision-making. The Internet of Things (IoT) connects devices for analysis and decision-making automation, while the IoMT is the network of interconnected medical devices for real-time monitoring and decision support in healthcare focuses on integrating medical devices to serve those who cannot easily access medical services, such as those in rural areas or the elderly [8]. The IoMT platform can collect medical attributes directly from patients using sensors or a user interface and use AI-based techniques to provide proper guidelines for patients and assist medical experts in timely treatment. In countries with insufficient medical infrastructure, such as during the COVID-19 pandemic, IoMT could have been used to provide healthcare support to remote or needy individuals, potentially saving many lives. An automatic system for epidemic virus estimated and early alarm was developed and will be discussed in the following sections of the paper, Section 13.2 covers related research on the application of AI in Image Classification for the Internet of Medical Things (IoMT). Section 13.3 presents the proposed design of an IoMT system specifically for pandemic diseases. Section 13.4 addresses a significant problem. Section 13.5 outlines the methodology employed in this study. The results of the image classification experiments are presented in detail in Section 13.6. Section 13.7 gives a conclusion for this study.

13.2 RELATED WORK

The COVID-19 pandemic has resulted in a large number of deaths and illnesses worldwide, and the treatment of lung infections caused by the virus presents significant challenges. While CT or X-ray images can assist in diagnosis and treatment selection, radiologists' interpretation may be influenced by bias or inexperience [9]. The (IoMT) aims to program health provisions and provide exact and economical care, particularly for distant patients, by integrating multiple medical devices for constant health observation. Measuring devices are implanted in the human being to measure various health check-up attributes, and treatment support is provided based on the measured values with the assistance of AI techniques and the advancement of internet [10]. IoMT handles a vast number of medical records, which are stored in the cloud. Various medical applications utilize IoMT, including the development of a profitable and exact IoMT building block for disruptive sleep apnea, where an SpO_2 sensor continuously measures blood oxygen levels and rate of the heart. The measured data is interconnected to the cloud for widespread health evaluation, and remote patients receive warnings through a mobile app based on the analysis report [11,12].

The IoMT presents a great opportunity to improve healthcare by enhancing health supervision, management, and procedures, leading to better

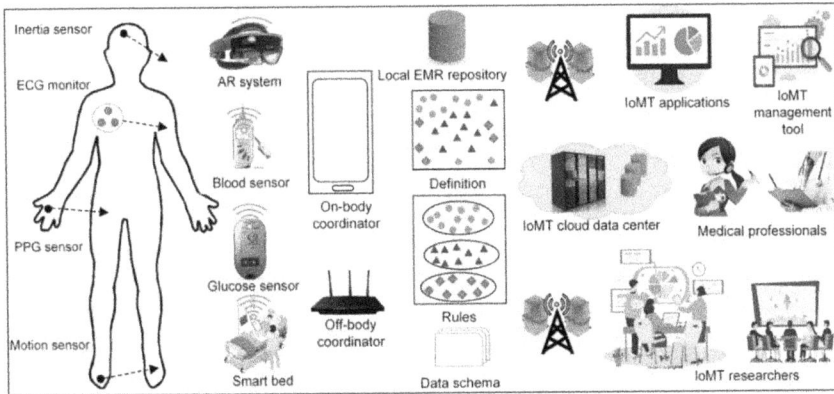

Figure 13.1 The architecture of IoMT shows the complete lifecycle [4].

human well-being and quality of life. The predicted shortage of healthcare professionals by the World Health Organization (WHO) by 2030 has spurred the development of cost-effective smart healthcare infrastructure and solutions to prevent diseases, reduce healthcare costs, and improve human life quality. The projected 31% increase in the world population by 2050, with 16% being waged over, will result in a rise in chronic diseases, which account for a significant proportion of deaths and global healthcare spending. While several terms have been used to refer to IoMT, it remains the most commonly used term in academia and research. The data sensitivity of IoMT devices, which contain patients' sensitive data, makes data protection a top priority [4].

The IoMT relies on the seamless flow of vital information between individuals and objects through the use of Cyber-Physical Systems (CPS), hardware, software, and communication devices. Through the utilization of ubiquitous and pervasive computing, IoMT devices can gather crucial data that can be employed for the autonomous management of healthcare infrastructure. As can be seen in Figure 13.1 architecture of IoMT shows the complete life cycle.

13.3 LOW POWER WIDE AREA NETWORK

Low Power Wide Area Network (LPWAN) is a kind of wireless network designed to enable wide-range communication between devices with low data rates and low power utilization. LPWAN technology is becoming increasingly popular in the IoT and machine-to-machine (M2M) communication domains because of its low cost, low power consumption, and long-range capabilities [13]. LPWAN technology offers a cost-effective, low-power solution for IoT applications that require long-range

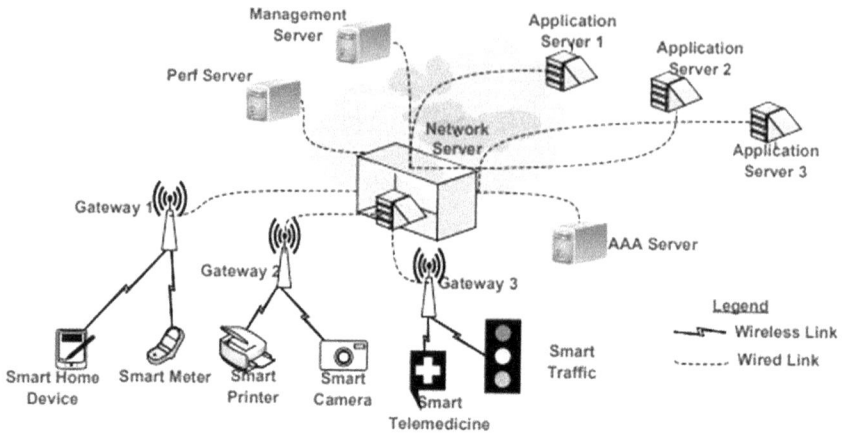

Figure 13.2 Basic LPWAN structure.

communication and low data rates. Figure 13.2 illustrates the essential framework of an LPWAN. As the IoT market continues to grow, LPWAN technology is expected to become increasingly popular due to its scalability, efficiency, and long-range capabilities [14].

13.3.1 IoMT in LPWAN

Medical sensors and connected devices that deploy low-power, long-range networks are used in IoMT in LPWAN. As a result, telemedicine, predictive maintenance, asset tracking, environmental monitoring, and real-time patient monitoring and data collection are all made possible. IoMT applications frequently leverage LPWAN technologies like LoRa and Sigfox because of their long-range capabilities, low power requirements, and affordable infrastructure. Below Table 13.1 shows the description of IoMT in LPWAN. However, it has the potential to transform healthcare by enhancing patient outcomes, lowering costs, and boosting productivity.

13.3.2 LPWAN technologies

LPWAN standards that are now in use include Sigfox, LoRa, Weightless, DASH7, NB-IoT, and Ingenu-RPMA, to name just a few. These standards are unique in that they address common IoT issues such as scalability, security, and power consumption. Each LPWAN standard also has particular technological requirements, such as medium access control (MAC) and physical layer standards [15]. Table 13.2 illustrates LPWAN technologies' description with their frequency, range, and data range. Therefore, it's important to understand the individual characteristics of each LPWAN

Table 13.1 IoMT in LPWAN for healthcare

Application	Description
Remote patient monitoring	Monitoring conditions of the patient, medication compliance, and other health data remotely from a patient's home or another non-clinical context using medical devices with LPWAN capabilities. This enables ongoing, real-time monitoring and can aid in the early detection of any potential health issues.
Asset tracking	Medical equipment and supplies are monitored via LPWAN-capable tags, which minimizes the risk of theft or misuse and improves operational efficiency.
Environmental monitoring	To provide the best conditions for patients and equipment, healthcare facilities should use LPWAN-capable sensors to monitor temperature, humidity, and other environmental variables.
Predictive maintenance	Predicting when maintenance is necessary for equipment, minimizing downtime, and increasing productivity using data from LPWAN-enabled medical devices.
Telemedicine	LPWAN-enabled communication technologies are used to provide remote medical consultations and treatment, allowing patients to get care from any location and minimizing requiring in-person visits.

standard as they may vary from one another. In summary, there are several LPWAN technologies with different technical specifications that cater to specific IoT requirements [13].

Study on IoMT work is given in Table 13.3. To address the shortage of medical infrastructure during the COVID-19 pandemic, particularly for orthopedic patients who are unable to attend routine checkups, the authors have devised an IoMT system [22]. This method enables telemedicine and cloud-based services to provide patients with remote medical care. An IoMT system for clinical nursing, in which temperature and pulse rate sensors are affixed to the patient's body, has been proposed by another study [23]. By allowing clinical nurses to monitor patients remotely while connected to the internet and cloud technology, the strain on healthcare personnel is reduced. IoMT, as a whole, consists of sensing components for gathering medical data from patients and processing components for making judgments automatically using the data collected [24].

Table 13.2 LPWAN technologies description

LPWAN Technology	Modulation Technique	Frequency	Range	Data Rate	Power Consumption	License Type
LoRaWAN	LoRa	Sub-GHz	Up to 10 km	0.3–50 Kbps	Low	Open
Sigfox	Ultra-Narrowband	Sub-GHz	Up to 40 km	Up to 100 bps	Ultra-Low	Proprietary
NB-IoT	Cellular	Licensed Spectrum	Up to 10 km	25–250 Kbps	Low	Licensed
LTE-M	Cellular	Licensed Spectrum	Up to 10 km	Up to 1 Mbps	Low	Licensed
Weightless	Various	Sub-GHz and TV White Spaces	Up to 10 km	Up to 100 Kbps	Low	Open

Table 13.3 Summary of IoMT work

Title	Techniques Used	Strength
FogCloud ArchitectureDriven Internet of Medical Things Framework for Healthcare Monitoring [16]	Deep learning, IoMT, Fog computing, Cloud computing	Early detection of COVID-19, improves patient outcomes, reduces spread of infection, potential for real-time monitoring
Parkinson's Disease Management via Wearable Sensors: A Systematic Review [17]	Wearable sensors, IoT, Machine learning	Non-invasive, continuous monitoring, early detection of Parkinson's disease, potential for remote monitoring and management
A Healthcare Monitoring System for the Diagnosis of Heart Disease in the IoMT Cloud Environment [18]	IoT, Machine learning, Heart rate variability analysis	Early prediction of heart disease, potential for personalized treatment plans, non-invasive monitoring
IoT Adoption and Application for Smart Healthcare: A Systematic Review [19]	IoT, Systematic review	Comprehensive overview of IoT applications in healthcare, potential for identifying research gaps and future directions
A Survey of IoT-Based Fall Detection for Aiding Elderly Care: Sensors, Methods, Challenges and Future Trends [20]	IoT, Machine learning, Fall detection algorithms	Early detection of falls, potential for reducing injuries and healthcare costs, non-invasive monitoring
An Intelligent Real Time IoT-Based System (IRTBS) for Monitoring ICU Patient [21]	IoT, Real-time monitoring, Predictive analytics	Early detection of patient deterioration, potential for personalized treatment plans, improves patient outcomes

13.3.3 Deep learning model for classification

Deep learning model for classification refers to the use of artificial neural networks with multiple hidden layers to train a model that can classify data into different categories or classes. The deep learning model uses a large number of training examples to learn the underlying patterns in the data and can make accurate predictions on new, unseen data.

Deep learning models for classification have been successfully used in various applications such as image recognition, speech recognition, natural language processing, and more. These models are very good at tasks like object recognition, sentiment analysis, and fraud detection because they can learn intricate correlations between input characteristics and output classes. The IoMT-enabled deep CNN in the fog-cloud network is shown in Figure 13.3. Several layers of neurons make up the architecture of a

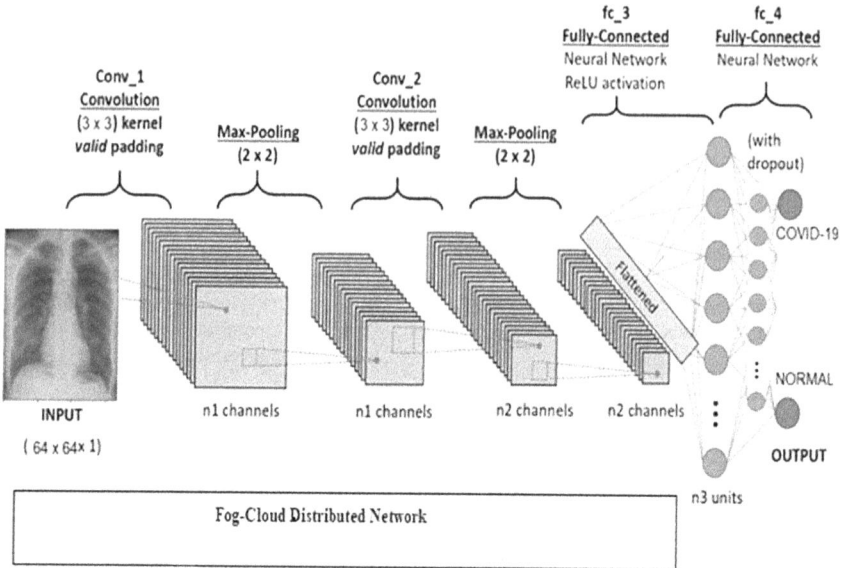

Figure 13.3 IoMT-enabled deep convolutional neural network in Fog-Cloud Network.

deep learning model for classification, with each layer applying a particular operation to the input data. A number of hidden layers perform non-linear changes on the input layer's data before it is given back to the input layer. The output layer, the last layer in the model, generates the predicted class label based on the input data. A deep learning classification model is trained by minimizing a loss function that measures the difference between the output that was predicted and the actual output. Backpropagation is a technique that does this by calculating the gradients of the loss function with respect to the model parameters and updating the parameters using gradient descent optimization. Yet, deep learning models for classification have transformed the area of AI recently and provide a potent tool for resolving challenging classification issues.

An artificial neural network that can discriminate between images was developed in this study using a CNN. Figure 13.3 illustrates the CNN model, which has two convolution layers and two max pooling layers. The convolution layer, which is the first layer of an artificial neural network, collects data from an input image while maintaining the relationships between its pixels. A 3D tensor with the dimensions of height, width, and channel is the output of each convolution layer and max pooling layer. More intricate structures from the original image can be reconstructed as more layers are applied. Convolution recovers characteristics like initial edges. The pooling layer reduces the size of the convolved feature map while retaining spatial invariance and requiring fewer CPU resources to analyze the input. The packed-down features matrix is sent to the fully linked layer, which has two dense layers for the classification result. The top layer contains 256 nodes, whereas the bottom layer contains just one node. The number of input images and the Processor have an impact on how many convolution layers and pooling layers are used.

13.3.4 Covid-19 data description

With the help of the Internet of Medical Things, it is possible to identify numerous unique characteristics of coronavirus sickness in order to establish an effective infection management and early detection system against COVID-19 [25]. A dataset including photographs of both sick and uninfected people must be gathered for this purpose [26]. These kinds of datasets are crucial for diagnosis in a fictitious situation when all practitioners and radiologists are afflicted, and no one is available to help effectively. From the COVID-19 image data collection, an open-source dataset, 130 sample chest X-ray training images and 18 sample chest X-ray test images were used in this work [27].

13.3.5 COVID-19 impact on IoMT

The healthcare industry has been severely impacted by the outbreak of COVID-19 and has undergone significant changes. To prevent infections and viral infection, end users are seeking counseling and treatment through digital media as healthcare facilities struggle with capacity issues [28]. These factors have resulted in an increased demand for IoMT. Additionally, the government is focusing on remote patient monitoring, and there is a growing demand for healthcare services, which has contributed to the expansion of the sector. Healthcare service providers have declared a rise in financing for connected medical research to address the issues brought on by the COVID-19 pandemic. According to a study, connected medical devices will account for roughly 48% of all medical devices in 2021 and approximately 68% of all medical devices in 2027. Manufacturers of linked devices predict that by 2025, expenditure on connected medical devices will have increased to around 42% from about 34% currently along with increasing their R&D expenses. These trends imply that IoMT is most likely to have a big impact on the healthcare industry in the future [4].

13.4 MAJOR PROBLEM

The COVID-19 pandemic has posed significant difficulties for the world's healthcare sector, notably in terms of quickly locating and diagnosing cases. For the virus to be contained, its effects on the populace to be reduced, and lives to be saved, early diagnosis is essential. Unfortunately, conventional diagnosis techniques like chest X-rays and RT-PCR testing need a lot of time and money, involving qualified medical personnel and specialized equipment [10]. It is also difficult to recognize and manage the condition because of a lack of medical infrastructure and the inability of many people to get frequent checkups [29]. As a potential option for the early diagnosis of COVID-19, the integration of IoMT and Fog-Cloud healthcare networks with deep learning techniques has been proposed [30]. The suggested method trains deep learning models that can recognize the presence of COVID-19 in patients using the enormous amount of data produced by IoMT devices such sensors, wearables, and medical imaging equipment. To provide real-time analysis and decision-making capabilities, the models can subsequently be implemented in Fog-Cloud healthcare networks. This can greatly increase the efficiency and precision of COVID-19 diagnosis.

As shown in Figure 13.4, the IoMT market is expected to grow from USD 82.324 billion in 2021 to USD 284.5 billion in 2027.

The problem statement of this study, therefore, is to develop a deep learning model capable of detecting COVID-19 in IoMT Fog-Cloud healthcare networks. The model must be able to process and analyze large amounts of medical data generated by IoMT devices and provide accurate and timely results to healthcare professionals. Table 13.4 shows the comparison of IoT and IoMT in cloud healthcare networks, this aim is to provide a more efficient and effective approach to COVID-19 diagnosis, especially in situations where traditional diagnosis methods are not readily available or are not feasible due to resource constraints. Ultimately, the goal is to improve patient outcomes, control the spread of the disease, and save lives.

Research Questions

1. How can the integration of deep learning algorithms improve the early detection of COVID-19 in IoMT Fog-Cloud Healthcare in LPWANT?

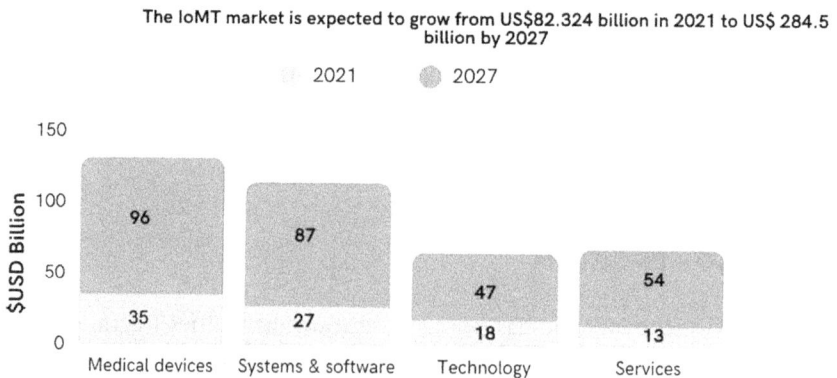

Figure 13.4 Forecasting the growth of the IoMT market.

Table 13.4 Comparison of IoT and IoMT in cloud healthcare networks

Features	IoT	IoMT
Definition	A network of physical devices, vehicles, home appliances, and other items that are embedded with electronics, software, sensors, and connectivity to enable these objects to connect and exchange data.	The Internet of Medical Things refers to a connected infrastructure of medical devices, software applications, and health systems and services that can communicate with various healthcare IT systems.
Healthcare Application	Remote monitoring, Telemedicine, Wearable health devices	Remote patient monitoring, Telemedicine, Patient Health Management, Wearable health devices
Data Security	Security risks are higher due to a large number of connected devices, and vulnerabilities in IoT devices.	Data security is critical, as patient information is highly confidential and can lead to privacy concerns. IoMT networks are designed to be highly secure and comply with HIPAA regulations.
Data Storage	Data is typically stored in cloud-based platforms.	Data can be stored in cloud-based platforms or local servers.
Network architecture	Typically follows a centralized architecture, where the data is collected from sensors and devices and sent to a central processing unit.	IoMT networks can be decentralized, centralized, or hybrid. A hybrid architecture allows for a balance between local and cloud-based processing.
Technology Stack	Wi-Fi, Bluetooth, Zigbee, Z-wave, Cellular Network	Wi-Fi, Bluetooth, Zigbee, Z-wave, Cellular Network
Benefits	Improved efficiency, reduced costs, and better patient outcomes.	Better health outcomes, reduced costs, personalized healthcare, and improved patient engagement.
Limitations	Security risks, privacy concerns, and compatibility issues between devices from different manufacturers.	High implementation costs, the need for highly skilled professionals, and interoperability issues between devices from different manufacturers.

2. What are the potential benefits and limitations of using deep learning algorithms for COVID-19 detection in IoMT Fog-Cloud Healthcare Network?

3. How can IoMT Fog-Cloud Healthcare Network be optimized to enhance the performance of deep learning algorithms for COVID-19 detection?

13.5 METHODOLOGY

The various steps and elements of a typical deep learning workflow created for the early detection and categorization of COVID-19 utilizing IoMT are described in this research study. The paper also looks at the drawbacks and difficulties of this process and assesses the most recent developments and fashions, with a focus on deep learning methods. Section 13.3 of this study's approach for doing research provides a detailed explanation. To give a thorough explanation of the suggested deep learning method for early detection, the workflow process is illustrated in Figure 13.5, covering the entire process from start to finish. Integration of IoMT fog-cloud can be seen in Figure 13.6.

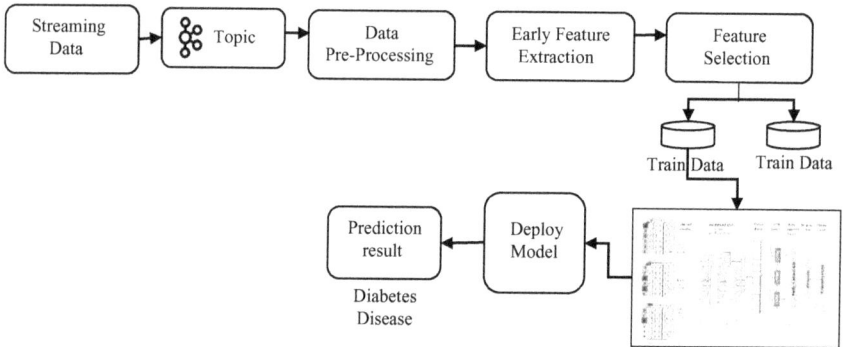

Figure 13.5 Proposed IoMT support system for COVID-19.

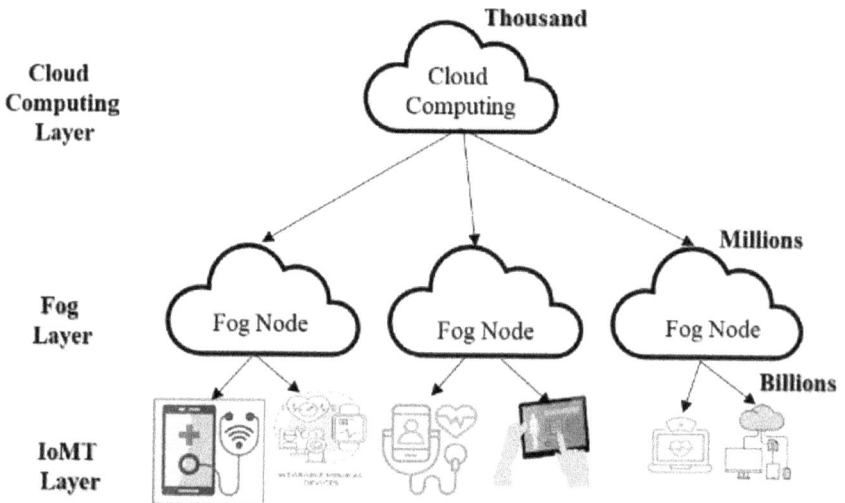

Figure 13.6 IoMT-fog-cloud integration.

13.5.1 Image preprocessing and data generation

With a focus on data augmentation, this part is addressed to image preparation using the Keras framework. To artificially create new training data from the existing training data in order to increase the training dataset's sample size, this method is known as data augmentation. For instance, this can entail altering the photos' object alignment, location, and direction.

13.5.2 Training the fully connected neural network

The study included 148 photos in total, of which 130 were used to train the model and the remaining 18 to test it. The testing images were similarly separated into two classes as the training images, which were each divided into two categories. The flattened output was sent to a feed-forward layer of a fully connected artificial neural network to process the images. Each training iteration used the backpropagation method, which raised the model's quality. The neural network's parameters become closer to values that result in higher accuracy with each iteration. During each epoch, the fully connected neural network improves the model's ability to distinguish between the features. The model can identify some dominant features in photos.

13.5.3 How can the inclusion of deep learning algorithms improve the IoMT Fog Cloud Healthcare Network's ability to detect COVID-19 early?

In the IoMT Fog-Cloud Healthcare Network, the use of deep learning algorithms can enhance COVID-19 early detection in a number of ways. First, deep learning algorithms are capable of analyzing huge amounts of medical data, such as electronic health records and medical pictures, to find patterns and correlations that might point to COVID-19 infection. This can assist medical professionals in swiftly identifying and diagnosing COVID-19 infections in individuals, even if they are asymptomatic.

Second, to identify distinguishing features and traits of the disease, deep learning algorithms can be trained on vast datasets of COVID-19 cases. This can aid in the improvement of COVID-19 diagnosis accuracy as well as the creation of efficient medications and vaccinations. Finally, the combination of deep learning algorithms with IoMT and Fog-Cloud technologies can enable real-time monitoring and analysis of patient data, including vital signs, symptoms, and illness development. By enabling healthcare providers to quickly identify and react to changes in a patient's state, this may improve patient outcomes and reduce the spread of COVID-19 characteristics.

Improved accuracy: By examining a large amount of data and spotting tiny patterns that a human analyst might miss, the incorporation of deep learning algorithms might increase the COVID-19 detection's precision.

Moreover, it can lessen the incidence of both false positives and false negatives.

Faster detection: The time required for COVID-19 identification is decreased by deep learning algorithms' ability to analyze vast amounts of data quickly and deliver speedy findings. This can aid in the prompt detection and treatment of patients who are infected.

Remote detection: Remote detection of COVID-19 may be made possible by integrating deep learning algorithms with the IoMT Fog-Cloud Healthcare Network. This lowers the risk of transmission to healthcare workers and allows for remote diagnosis and monitoring of patients.

Scalability: Deep learning techniques are scalable to handle massive volumes of data, enabling real-time analysis of data from many sources. This can aid in the quick identification and monitoring of outbreaks.

Cost-effective: By the automation of the analytical process and the elimination of expensive diagnostic tools and manual labor, the incorporation of deep learning algorithms can lower the cost of COVID-19 detection.

By integrating deep learning techniques, COVID-19 detection on the IoMT Fog-Cloud Healthcare Network can be made more precise, swift, scalable, and economical. Moreover, it may enable distant outbreak detection and surveillance, improving the effectiveness and efficiency of pandemic management.

13.5.4 What are the benefits and potential drawbacks of using deep learning techniques for COVID-19 identification in the IoMT Fog-Cloud Healthcare Network?

In the IoMT Fog-Cloud Healthcare Network, deep learning algorithms can significantly contribute to the early diagnosis of COVID-19. Deep learning models can learn to recognize patterns and traits that are symptomatic of infection by making use of the enormous volumes of data provided by the IoT-enabled medical equipment. This may result in a quicker and more precise diagnosis, which in turn may aid in stopping the virus's spread. The adoption of deep learning methods for COVID-19 identification in the IoMT Fog-Cloud Healthcare Network may have certain drawbacks, though. The necessity for a significant amount of high-quality training data, the possibility of bias and overfitting, and the difficulties in interpreting the outcomes of deep learning models are a few of these. IoMT devices may transfer sensitive patient data over the internet, hence there may also be privacy and security issues with their use in the healthcare industry [11].

Potential benefits

Increased accuracy: Deep learning algorithms have the capacity to uncover intricate patterns in data, improving the COVID-19 detection efficiency.

This may lessen false negatives and false positives, improving patient outcomes in the process

Early detection: To personalize treatment approaches and identify risk factors, deep learning algorithms can examine patient data. In addition to lowering healthcare expenditures, this can enhance patient outcomes.

Personalized medicine: To personalize treatment approaches and identify risk factors, deep learning algorithms can examine patient data. This can enhance patient outcomes and lower medical expenses.

Remote monitoring: Remotely monitoring patients and spotting COVID-19 signs before they worsen can be done using deep learning algorithms. This can lessen the need for hospital stays and enhance patient outcomes.

Potential limitations

Data bias: Deep learning systems rely on a significant amount of high-quality data, and biased data can produce unreliable findings. Disparities in healthcare outcomes may result from this.

Model complexity: The complex nature of deep learning algorithms can be challenging to comprehend how they get to their conclusions. This may make it difficult to evaluate the findings and create distrust toward technology.

Limited interpretability: Understanding how deep learning algorithms come to their findings can be tricky since they can be challenging to interpret. Health professionals may find it challenging to rely on technology and make wise decisions as a result.

Data privacy and security: Large patient data sets are needed for deep learning algorithms, which raises questions regarding data security and privacy. Patient data misuse and breaches may result from improper data security.

13.5.5 How can the IoMT Fog-Cloud Healthcare Network be optimized to enhance how well COVID-19 detection deep learning algorithms perform?

To increase the effectiveness and efficiency of diagnosis for COVID-19 patients, it is crucial to optimize IoMT Fog-Cloud Healthcare LPWAN for deep learning algorithms for COVID-19 detection. A network of connected medical equipment, sensors, and cloud-based servers called the IoMT Fog-Cloud Healthcare LPWAN can be used to gather and send medical data for analysis. The performance of deep learning algorithms for COVID-19 detection can be enhanced by optimizing this network, which can aid in the early diagnosis and treatment of COVID-19 patients [31].

The IoMT Fog-Cloud Healthcare LPWAN can be optimized in a number of ways, including:

Improving data quality: Deep learning algorithms work best when given high-quality data. The IoMT Fog-Cloud Healthcare LPWAN can be configured to guarantee that the information gathered from medical devices and sensors is precise, dependable, and devoid of noise or artifacts that could obstruct analysis.

Increasing data volume: large amounts of data are needed for deep learning algorithms to accurately train. We can gather and transmit more data from medical devices and sensors by enhancing the IoMT Fog-Cloud Healthcare LPWAN, which will increase the precision and dependability of the deep learning models.

Reducing latency: Deep learning algorithms' performance can be greatly impacted by latency. We can speed up data transmission between medical devices, sensors, and cloud-based servers by enhancing the IoMT Fog-Cloud Healthcare LPWAN. This will increase the efficiency and precision of Covid-19 detection.

Enhancing computational power: The training and optimization of deep learning algorithms require a lot of processing resources. We may improve the performance of the deep learning models by increasing the computational power that is accessible to them by optimizing the cloud-based servers in the IoMT Fog-Cloud Healthcare LPWAN.

13.5.6 IoMT for cloud side

IoMT, or the "Internet of Medical Things," is a network of networked medical technology and sensors that gathers and sends patient health information to the cloud or other distant servers for processing and archiving. The cloud computing infrastructure that makes it possible to store, process, and analyze the enormous amount of data produced by IoMT devices is referred to as the "cloud side" of IoMT. Healthcare providers can remotely monitor and diagnose patients, ease the strain on hospital resources, and enhance the standard of care by utilizing the cloud's capabilities. Because it offers the processing power and storage required for training and operating deep learning algorithms, cloud computing also makes it possible to utilize these algorithms for COVID-19 detection and other medical applications with complex models [17,32]. Figure 13.7 shows IoMT framework for cloud side.

13.6 SIMULATION RESULTS

The COVID-19 pandemic has brought attention to the value of early virus identification and diagnosis. As the COVID-19 virus binds its spikey surface

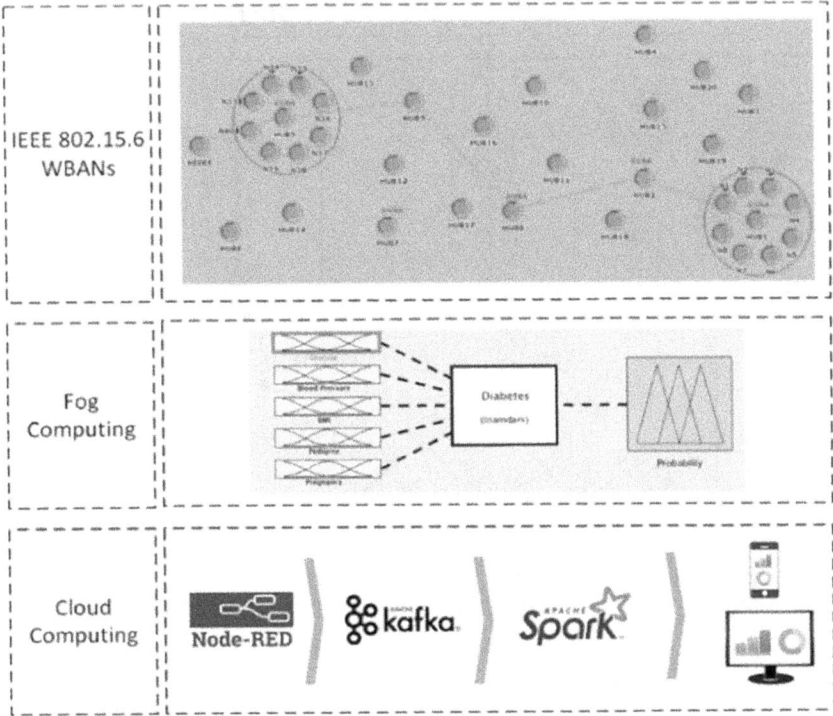

Figure 13.7 IoMT framework for cloud side.

proteins to receptors on healthy cells in the lungs, chest X-rays are one of the most often utilized diagnostic methods to detect lung abnormalities associated with the virus. Yet, as the number of COVID-19 cases continues to climb globally, medical practitioners have found it more challenging to examine and diagnose chest X-rays.

With the ultimate goal of early COVID-19 sickness detection, a deep learning model utilizing a CNN was developed to differentiate between infected and uninfected patients' chest X-ray images. The CNN model was trained using a dataset of chest X-ray images from 130 patients, and it was completed with an astounding accuracy level of 0.98 and a loss of 0.04. The accuracy-loss evolution data, shown in Figures 13.8 and 13.9, show the model's ability to accurately classify patients' chest X-ray images as either infected or uninfected. It is critical that the deep learning model be able to identify anomalies related to COVID-19 in chest X-ray images for the aim of early detection and diagnosis of the virus. The deep learning algorithm may help medical practitioners swiftly detect COVID-19 cases because it can diagnose billions of people in a single day, which can be critical in preventing the spread of the virus.

Accuracy over 100 Epochs

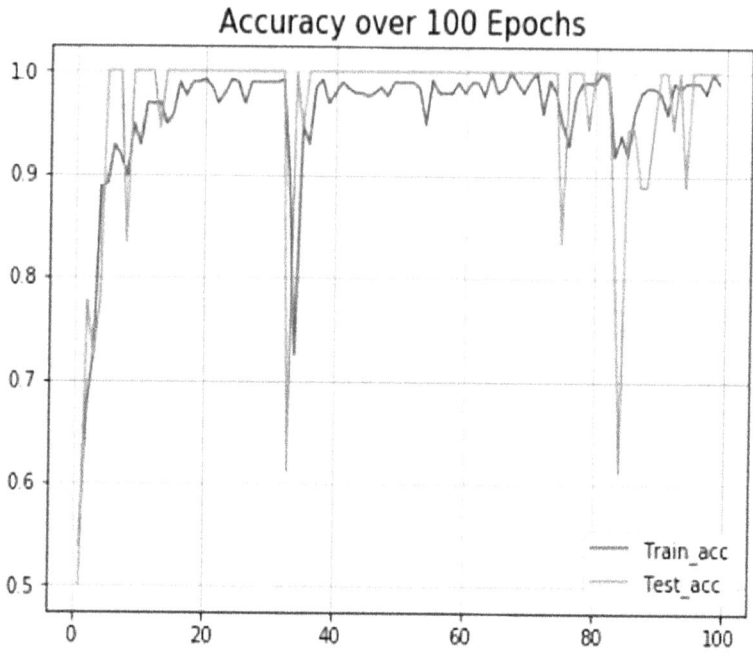

Figure 13.8 Accuracy-over graph.

Loss over 100 Epochs

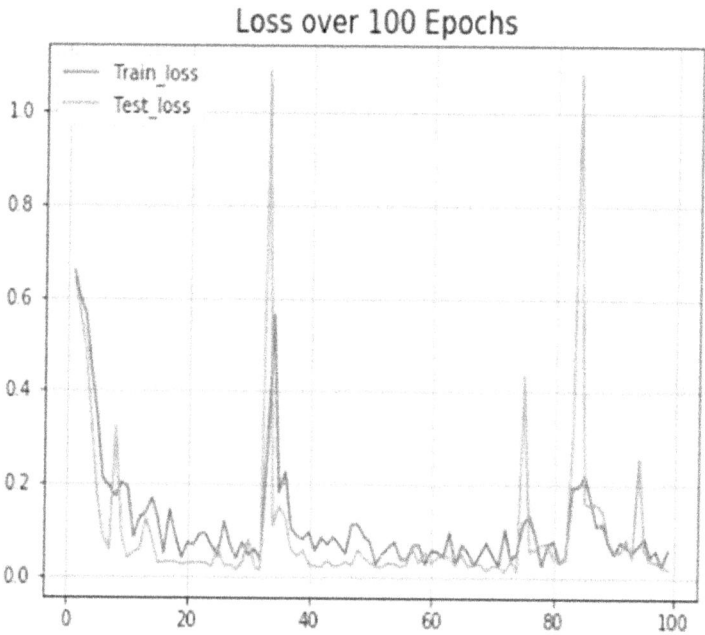

Figure 13.9 Accuracy-over and accuracy-loss graph.

The random sample size of the dataset used to train the CNN model is one of the study's shortcomings, though. In the future, more data points can be utilized to train the model, and the CNN model's fundamental structure can be changed to enable deeper and more precise analysis of chest X-ray images. In addition, a graphical user interface will be created to make the deep learning model accessible to physicians and radiologists at clinics and hospitals. Notwithstanding these drawbacks, the deep learning model created in this study has demonstrated excellent results in properly recognizing lung abnormalities associated with COVID-19 in chest X-ray pictures, which can help in the early diagnosis and treatment of COVID-19 cases. Technology advancements, like the CNN model created in this study, offer hope for improved early detection and control of the virus as the COVID-19 pandemic continues to affect communities around the world. Figures 13.8 and 13.9 show the implementation of the accuracy over the loss graph on 100 epochs and the accuracy over graph on 100 epochs, respectively.

Figure 13.10 illustrates how a deep learning model was tested in the study utilizing single-prediction data images. The model correctly identified the patient in the left-hand chest X-ray image as having COVID-19. The model successfully identified the healthy patient in the image on the right as being uninfected, in contrast. This indicates the model's capability to discriminate between normal lung pictures and COVID-19-related anomalies with accuracy. The COVID-19 cases and normal lung images successfully classified suggest that the deep learning model has the potential to be improved and employed as a useful diagnostics application. The model has a high level of accuracy and can swiftly evaluate vast amounts of data, making it a useful tool for medical professionals to identify COVID-19 cases quickly and reliably, depicts anticipated COVID-19 and normal categorization, demonstrating how important timely detection is for controlling the virus's transmission. With regard to using technology to enhance illness diagnosis and

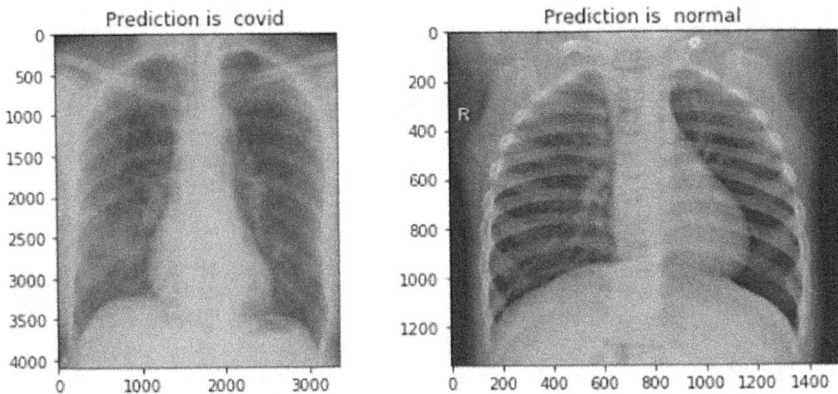

Figure 13.10 Predicted COVID-19 and normal classification.

control during the COVID-19 pandemic, these findings indicate a significant advancement. Medical personnel may be better prepared to respond to COVID-19 and other infectious diseases by utilizing the deep learning model's capabilities.

13.7 CONCLUSION

In conclusion, a tool that can precisely forecast the likelihood of infection could revolutionize the effort to stop epidemics and stop the spread of illness. Early detection and diagnosis are crucial, as the COVID-19 pandemic has proven, and deep learning models have shown considerable potential in this area. The opinions of radiologists and doctors must be considered when testing such applications. These experts can offer helpful insights to ensure the correctness and efficacy of the deep learning model because they have a plethora of experience in making medical diagnoses. Moreover, protocols must be established to guarantee that the application's outputs are properly evaluated, and medical practitioners should be trained in its use.

The results of this study point to the feasibility of developing a deep learning model that can reliably distinguish between healthy and infected chest X-ray images as a strategy for the early detection and diagnosis of COVID-19. The model's high level of accuracy and ability to quickly evaluate enormous volumes of data make it a valuable tool in the war against the virus.

It is crucial to remember that the study's breadth was constrained because the model was trained using only 130 patient data points. The dataset should be increased, and the model's structure should be improved in subsequent studies to enable more thorough picture analysis. To make the method easier for doctors and radiologists to use, a graphical user interface should also be created. Our capacity to identify and manage infectious diseases like COVID-19 can be considerably improved by utilizing the power of deep learning models and technologies. We can continue to create and improve these tools by working with medical professionals and spending money on research and development, ultimately saving lives and preserving public health.

REFERENCES

[1] N. Chen et al., "Epidemiological and clinical characteristics of 99 cases of 2019 novel coronavirus pneumonia in Wuhan, China: a descriptive study." *Lancet*, vol. 395, no. 10223, pp. 507–513, 2020, doi: 10.1016/S0140-6736(20)30211-7.

[2] D. Huang, Clinical features of severe patients infected with 2019 novel coronavirus_ a systematic review and meta-analysis. https://www.ncbi.nlm.nih.gov/pmc/articles/PMC7290556/

[3] A. A. Sathio, A. R. Lakhan, and T. Karachi, "Dynamic content enabled microservice for business applications in distributed cloudlet cloud network." *Int. J. Emerg. Trends Eng. Res.*, vol. 9, no. 7, pp. 1035–1039, 2021, doi: 10.30534/ijeter/2021/32972021.

[4] R. U. Rasool, H. F. Ahmad, W. Rafique, A. Qayyum, and J. Qadir, "Security and privacy of internet of medical things: A contemporary review in the age of surveillance, botnets, and adversarial ML." *J. Netw. Comput. Appl.*, vol. 201, no. January, p. 103332, 2022, doi: 10.1016/j.jnca.2022.103332.

[5] T. Shakeel et al., *A Survey on COVID-19 Impact in the Healthcare Domain: Worldwide Market Implementation, Applications, Security and Privacy Issues, Challenges and Future Prospects*, vol. 9, no. 1. Springer International Publishing, 2023. doi: 10.1007/s40747-022-00767-w."tailored.pdf."

[6] A. Naidu, "Factors affecting patient satisfaction and healthcare quality." International journal of health care quality assurance, vol. 22, no. 4, pp. 366–381, 2009, https://doi.org/10.1108/09526860910964834

[7] A. Tasdelen and B. Sen, "A hybrid CNN-LSTM model for pre-miRNA classification." *Sci. Rep.*, vol. 11, no. 1, pp. 1–9, 2021, doi: 10.1038/s41598-021-93656-0.

[8] E. Click, "Out of space." *Lancet Psychiatry*, vol. 2, no. 9, p. 765, 2015, doi: 10.1016/S2215-0366(15)00367-3.

[9] N. S. Abouzakhar, A. Jones, and O. Angelopoulou, "Internet of Things Security: A Review of Risks and Threats to Healthcare Sector." *Proc. -2017 IEEE Int. Conf. Internet Things, IEEE Green Comput. Commun. IEEE Cyber, Phys. Soc. Comput. IEEE Smart Data, iThings-GreenCom-CPSCom-SmartData 2017*, vol. 2018-January, no. February 2018, pp. 373–378, 2018, doi: 10.1109/iThings-GreenCom-CPSCom-SmartData.2017.62.

[10] N. Younes, Challenges in Laboratory Diagnosis of the Novel Coronavirus SARS-CoV-2. https://pubmed.ncbi.nlm.nih.gov/32466458/

[11] M. Al-rawashdeh, P. Keikhosrokiani, B. Belaton, M. Alawida, and A. Zwiri, "IoT adoption and application for smart healthcare: A systematic review." *Sensors*, vol. 22, no. 14, pp. 1–28, 2022, doi: 10.3390/s22145377.

[12] S. Thandapani et al., "IoMT with deep CNN: AI-based intelligent support system for pandemic diseases." *Electron.*, vol. 12, no. 2, 2023, doi: 10.3390/electronics12020424.

[13] U. Raza, P. Kulkarni, M. Sooriyabandara, "Low power wide area networks: An overview." *IEEE Commun. Surv. Tutor.* vol. 19, pp. 855–873, 2017.

[14] S. Farrell. LPWAN Overview; Technical Report; Internet Engineering Task Force (IETF); 2018. Available online: https://tools.ietf.org/id/draft-ietf-lpwan-overview-09.html (accessed on 5 November 2020).

[15] R. Yasmin, J. Petäjäjärvi, K. Mikhaylov, and A. Pouttu, "On the Integration of LoRaWAN with the 5G Test Network." In *Proceedings of the IEEE 28th Annual International Symposium on Personal, Indoor, and Mobile Radio Communications (PIMRC)*, Montreal, QC, Canada, 8 October 2017; pp. 1–6.

[16] E. Yıldırım, M. Cicioğlu, and A. Çalhan, "Fog-cloud architecture-driven internet of medical things framework for healthcare monitoring." *Med. Biol. Eng. Comput.*, vol. 61, no. 5, pp. 1133–1147, 2023, doi: 10.1007/ s11517-023-02776-4.

[17] H. Mughal, A. R. Javed, M. Rizwan, A. S. Almadhor, & N. Kryvinska. Parkinson's disease management via wearable sensors: a systematic review. IEEE Access, vol. 10, pp. 35219–35237, 2022.

[18] M. A. Khan & F. Algarni. "A healthcare monitoring system for the diagnosis of heart disease in the IoMT cloud environment using MSSO-ANFIS." *IEEE Access*, vol. 8, pp. 122259–122269, 2020.

[19] Al-Rawashdeh, M., Keikhosrokiani, P., Belaton, B., Alawida, M., & Zwiri, A. "IoT adoption and application for smart healthcare: A systematic review." *Sensors*, vol. 22, no. 14, p. 5377, 2022.

[20] M. E. Karar, H. I. Shehata, & O. Reyad. "A survey of IoT-based fall detection for aiding elderly care: sensors, methods, challenges, and future trends." *Appl. Sci.*, vol. 12, no. 7, p. 3276, 2022.

[21] B. Prajapati, S. Parikh, & J. Patel. "An Intelligent Real Time IoT Based System (IRTBS) for Monitoring ICU Patients." In *Information and Communication Technology for Intelligent Systems (ICTIS 2017)-Volume 2 2* (pp. 390–396). Springer International Publishing, 2018.

[22] M. Abomhara and G. M. Koien, "Security and Privacy in the Internet of Things: Current Status and Open Issues." *2014 Int. Conf. Priv. Secur. Mob. Syst. Prism. 2014- Co-located with Glob. Wirel. Summit*, no. January 2020, pp. 1–8, 2014, doi: 10.1109/PRISMS.2014.6970594.

[23] A. Aljumah, "Assessment of machine learning techniques in IoT-based architecture for the monitoring and prediction of COVID-19." *Electron.*, vol. 10, no. 15, 2021, doi: 10.3390/electronics10151834.

[24] S. Hamid, N. Z. Bawany, A. H. Sodhro, A. Lakhan, and S. Ahmed, "A systematic review and IoMT based big data framework for COVID-19 prevention and detection." *Electron.*, vol. 11, no. 17, 2022, doi: 10.3390/ electronics11172777.

[25] A. A. Toor, M. Usman, F. Younas, A. C. M. Fong, S. A. Khan, and S. Fong, "Mining massive e-health data streams for IoMT enabled healthcare systems." *Sensors (Switzerland)*, vol. 20, no. 7, pp. 1–24, 2020, doi: 10.3390/ s20072131.

[26] M. E. Karar, H. I. Shehata, and O. Reyad, "A survey of IoT-based fall detection for aiding elderly care: Sensors, methods, challenges and future trends." *Appl. Sci.*, vol. 12, no. 7, p. 3276, 2022, doi: 10.3390/app12073276.

[27] L. Wang, Z. Q. Lin, and A. Wong, "COVID-Net: A tailored deep convolutional neural network design for detection of COVID-19 cases from chest X-ray images." *Sci. Rep.*, vol. 10, no. 1, pp. 1–12, 2020, doi: 10.1038/ s41598-020-76550-z.

[28] A. Hemmati and A. M. Rahmani, "Internet of medical things in the COVID-19 era: A systematic literature review." Sustainability, vol. *14*, p. 12637, 2022.

[29] M. A. Khan and F. Algarni, "A Healthcare monitoring system for the diagnosis of heart disease in the IoMT cloud environment using MSSO-ANFIS." *IEEE Access*, vol. 8, pp. 122259–122269, 2020, doi: 10.1109/ ACCESS.2020.3006424.

[30] J. R. Joshi and N. Adhikari, "An overview on common organic solvents and their toxicity." *J Phrama. Res. Int.*, vol. 28, 1–18. [Online]. Available: https://click.endnote.com/viewer?doi=10.9734%2Fjpri%2 F2019&token=WzM0MTczNzcsIjEwLjk3MzQvanByaS8yMDE5Il0. T2E_bAcDOaLZO6FYXU7B9vEm3XI

[31] H. Mughal, A. R. Javed, M. Rizwan, A. S. Almadhor, and N. Kryvinska, "Parkinson's disease management via wearable sensors: A systematic review." *IEEE Access*, vol. 10, pp. 35219–35237, 2022, doi: 10.1109/ ACCESS.2022.3162844.

[32] A. Tasdelen, B. Sen, A hybrid CNN-LSTM model for pre-miRNA classification. *Sci. Rep.*, 11, 14125, 2021. doi: 10.1038/s41598-021-93656-0.

Index

For Product Safety Concerns and Information please contact our EU
representative GPSR@taylorandfrancis.com
Taylor & Francis Verlag GmbH, Kaufingerstraße 24, 80331 München, Germany